The Receptors

Volume 30

Series Editor
Giuseppe di Giovanni
Department of Physiology & Biochemistry
Faculty of Medicine and Surgery
University of Malta, Malta, Italy

The Receptors book Series, founded in the 1980s, is a broad-based and well-respected series on all aspects of receptor neurophysiology. The series presents published volumes that comprehensively review neural receptors for a specific hormone or neurotransmitter by invited leading specialists. Particular attention is paid to in-depth studies of receptors' role in health and neuropathological processes. Recent volumes in the series cover chemical, physical, modeling, biological, pharmacological, anatomical aspects and drug discovery regarding different receptors. All books in this series have, with a rigorous editing, a strong reference value and provide essential up-to-date resources for neuroscience researchers, lecturers, students and pharmaceutical research.

More information about this series at http://www.springer.com/series/7668

Kenji Hashimoto
Editor

The NMDA Receptors

Humana Press

Editor
Kenji Hashimoto
Division of Clinical Neuroscience
Chiba University Center for Forensic Mental Health
Chiba, Japan

The Receptors
ISBN 978-3-319-49793-8 ISBN 978-3-319-49795-2 (eBook)
DOI 10.1007/978-3-319-49795-2

Library of Congress Control Number: 2016962311

Printed on acid-free paper

This Humana Press imprint is published by Springer Nature
The registered company is Springer International Publishing AG
The registered company address is: Gewerbestrasse 11, 6330 Cham, Switzerland

History of NMDA Receptors

The research field of the excitatory amino acids, including L-glutamate and L-aspartate, had its origin in the discovery by Hayashi (1954) of the convulsive effects of these amino acids in the mammalian brain. Today, L-glutamate is widely accepted as the predominant excitatory neurotransmitter in the mammalian central nervous system (CNS), acting at a range of different glutamate receptor types, including *N*-methyl-D-aspartate (NMDA) receptors, α-amino-3-hydroxy-5-methylisoxazolepropionic acid (AMPA) receptors, and kainate receptors. The NMDA receptors are essential for neuronal development, synaptic plasticity, learning, and cell survival. In 1991, Nakanishi's group first cloned the NMDA receptor. In 2014, Furukawa's group and Gouaux's group reported the crystal structures of the NMDA receptors and structural models. The recent discovery indicates our understanding of the interdomain and intersubunit interactions that play key roles in NMDA receptor-mediated neurotransmission. Interestingly, the rapid antidepressant effects of the NMDA receptor antagonist ketamine in treatment-resistant patients with depression are the most important advance in 50 years. The role of the NMDA receptors in the CNS and peripheral functions is highlighted in the book.

The book, a result of the efforts of an international group of authors, has the aim of providing a history and an update of the functional status of the NMDA receptors, covering molecular, cellular, anatomical, biochemical, and behavioral aspects, to highlight its distinctive regulatory properties, the emerging functional significance, and the therapeutic potentiality in a number of diseases that are singled out in different chapters.

Kenji Hashimoto

Contents

Contributors

Milica Bozic Nephrology Department, Institute for Biomedical Research (IRB Lleida), Lleida, Spain

Shao-Rui Chen Department of Anesthesiology and Perioperative Medicine, Center for Neuroscience and Pain Research, The University of Texas MD Anderson Cancer Center, Houston, TX, USA

F. Errico Laboratory of Behavioural Neuroscience, Ceinge Biotecnologie Avanzate, Naples, Italy

Department of Molecular Medicine and Medical Biotechnology, University of Naples, Naples, Italy

Kenji Hashimoto Division of Clinical Neuroscience, Chiba University Center for Forensic Mental Health, Chiba, Japan

Eckhard Lammert, Ph.D. Institute of Metabolic Physiology, Heinrich Heine University Düsseldorf, Düsseldorf, Germany

Institute for Beta Cell Biology, German Diabetes Center, Leibniz Center for Diabetes Research Düsseldorf, Düsseldorf, Germany

Geoffroy Laumet Department of Anesthesiology and Perioperative Medicine, Center for Neuroscience and Pain Research, The University of Texas MD Anderson Cancer Center, Houston, TX, USA

Hisashi Mori Department of Molecular Neuroscience, Graduate School of Medicine and Pharmaceutical Sciences, University of Toyama, Toyama, Japan

Stéphane H.R. Oliet Neurocentre Magendie, Inserm U1215 and Université de Bordeaux, Bordeaux, France

Hui-Lin Pan Department of Anesthesiology and Perioperative Medicine, Center for Neuroscience and Pain Research, The University of Texas MD Anderson Cancer Center, Houston, TX, USA

Department of Anesthesiology and Perioperative Medicine, Unit 110, The University of Texas MD Anderson Cancer Center, Houston, TX, USA

Thomas Papouin Department of Neuroscience, Tufts University School of Medicine, Boston, MA, USA

Okka Scholz Institute of Metabolic Physiology, Heinrich Heine University Düsseldorf, Düsseldorf, Germany

Institute for Beta Cell Biology, German Diabetes Center, Leibniz Center for Diabetes Research Düsseldorf, Düsseldorf, Germany

A. Usiello Laboratory of Behavioural Neuroscience, Ceinge Biotecnologie Avanzate, Naples, Italy

Department of Environmental, Biological and Pharmaceutical Sciences and Technologies, The Second University of Naples (SUN), Caserta, Italy

José M. Valdivielso Nephrology Department, Institute for Biomedical Research (IRB Lleida), Lleida, Spain

Alena Welters, M.D., Ph.D. Institute of Metabolic Physiology, Heinrich Heine University Düsseldorf, Düsseldorf, Germany

Department of General Pediatrics, Neonatology and Pediatric Cardiology, University Children's Hospital Düsseldorf, Düsseldorf, Germany

Yukio Yoneda, Ph.D. Section of Prophylactic Pharmacology, Kanazawa University Venture Business Laboratory, Kanazawa, Ishikawa, Japan

Chapter 1
Overview of the NMDA Receptor

Hisashi Mori

Abstract The *N*-methyl-D-aspartate receptor (NMDAR) is a glutamate-gated ion channel that is critically involved in physiological and pathological functions in the central nervous system (CNS). Over the last 25 years, molecular biological studies revealed the molecular diversity of NMDAR subunits, the structural basis of NMDAR functions, and the *in vivo* functions of NMDAR subunits. Because NMDAR is involved in many diseases including neurodegenerative and psychiatric disorders, development of NMDAR-selective agonists and antagonists have great therapeutic potentials. In this chapter, I present an overview of the structure and function of NMDAR from molecular biological aspects.

Keywords *N*-methyl-D-aspartate receptors • GluN1 • GluN2 • GluN3 • Glutamate • D-Serine • Glycine • Neurodegenerative disorders • Psychiatric disorders • Gene knockout mice • Ketamine • Anti-NMDAR encephalitis

Abbreviations

AD	Alzheimer's disease
AMPA	α-Amino-3-hydroxy-5-methyl-4-isoxazole propionic acid
APV	D-2-Amino-5-phosphono-valerate
ATD	Amino-terminal domain
CNS	Central nervous system
CTD	Carboxy-terminal domain
GluR	Glutamate receptor
KO	Gene knockout
LBD	Ligand-binding domain

H. Mori (✉)
Department of Molecular Neuroscience, Graduate School of Medicine and Pharmaceutical Sciences, University of Toyama, Sugitani 2630, Toyama 930-0194, Japan
e-mail: hmori@med.u-toyama.ac.jp

K. Hashimoto (ed.), *The NMDA Receptors*, The Receptors 30,
DOI 10.1007/978-3-319-49795-2_1

LTP Long-term potentiation
NMDA *N*-methyl- D-aspartate
PSD Postsynaptic density

1.1 Introduction

Glutamate is one of the major excitatory neurotransmitters in the mammalian central nervous system (CNS). Glutamate binds to a specific receptor [glutamate receptor (GluR)], and induces excitatory neurotransmission and intracellular signal transduction. GluR is classified into the ionotropic (iGluR) and metabotropic (mGluR) types on the basis of their speed of neurotransmission and signaling mechanisms. iGluRs are further pharmacologically classified into three major subtypes, namely, α-amino-3-hydroxy-5-methyl-4-isoxazole propionic acid (AMPA), kainate, and *N*-methyl- D-aspartate (NMDA) receptor channels.

Historically, to analyze the role of acidic amino acids such as L-glutamate and L-aspartate in CNS, Watkins [1] synthesized a series of acidic amino acids and Curtis and Watkins [2] identified the agonist NMDA and the antagonist D-2-amino-5-phosphono-valerate (APV) to be highly selective to a GluR subpopulation. These specific ligands are used to identify the function of NMDAR and to discriminate the involvement of NMDA- and non-NMDA-type GluR in CNS functions. The unique properties of NMDAR are required for agonist (glutamate) and coagonist (glycine or D-serine) bindings, and release from voltage-dependent Mg^{2+} blocking for activation. One of the fundamental functions of CNS that depend on NMDAR activation is learning and memory. The hippocampus has an essential role in learning and memory. The use-dependent change and maintenance of synaptic transmission efficacy are the mechanisms underlying synaptic plasticity, which is considered to provide the physiological basis for information storage in the brain. One form of synaptic plasticity is enhancement of synaptic transmission called long-term potentiation (LTP). Experimentally, tetanic electrical stimulation-induced LTP in the hippocampal CA1 region has been studied as a model of an activity-dependent change in synaptic transmission efficacy [3]. Using APV, Collingridge et al. [4] identified the critical role of NMDARs in the induction of synaptic plasticity in the hippocampus. Subsequently, Morris et al. [5] showed the impairment of spatial learning of rats after the treatment of the hippocampus with APV. LTP and learning and memory are attenuated following the treatment with APV, indicating the critical role of NMDAR in these processes. Furthermore, many physiological and pathological roles of the NMDAR were examined using selective agonists and antagonists of NMDAR. In 1991, the first NMDAR subunit was identified by molecular biological approaches [6].

1.2 Molecular Diversity of NMDAR

1.2.1 Nomenclature of NMDAR Subunits

In this chapter, I use the names of NMDAR subunits on the basis of nomenclature of International Union of Basic and Clinical Pharmacology, namely, GluN1, GluN2A, GluN2B, GluN2C, GluN2D, GluN3A, and GluN3B. These subunits are also named on the basis of genetic nomenclature (for example, according to the nomenclature of Human Genome Organization) as GRIN1, GRIN2A, GRIN2B, GRIN2C, GRIN2D, GRIN3A, and GRIN3B.

1.2.2 Molecular Cloning and Heteromeric Nature of NMDAR

In 1991, the first NMDAR channel subunit GluN1 cDNA was functionally expressed and cloned using a *Xenopus* oocyte expression system [6]. From the retrospective view point, this cDNA cloning was pure luck because the GluN1 subunit is a glycine-binding subunit and the GluN1 subunit alone is not sufficient for the functional expression of NMDAR in mammalian cells. However, *Xenopus* oocytes express low levels of the endogenous NMDA-type glutamate-binding subunit *Xen*GluN2 [7]. Subsequently, eight splice variant forms of the GluN1 subunit were identified [8–10] (Fig. 1.1a). Because the NMDAR subunit family members have significant homology with other non-NMDAR-type GluR subfamilies, GluN2 members of the second NMDAR subfamily (GluN2A, GluN2B, GluN2C, and GluN2D) were identified by cDNA cross-hybridization, PCR-based homology cloning, and functional expression [11–15]. Combination with GluN1 and one of the members of GluN2 reconstitutes highly active NMDAR channels. These findings suggest the heteromeric nature of active NMDAR composed of GluN1 and GluN2 subfamilies. GluN3 subunit members of the third NMDAR subfamily (GluN3A and GluN3B) were cloned by PCR-based homology cloning [16–18]. Incorporation of GluN3 subunits into a heteromeric NMDAR composed of GluN1 and GluN2 suppresses the activity of NMDAR channels, suggesting the role of GluN3 as an inhibitory subunit of NMDAR. Heteromeric NMDAR channels composed of GluN2 (glutamate binding) and GluN3 (glycine binding) show no activity. The GluN1 and GluN3 heteromer shows the activity of glycine-gated ion channels in a *Xenopus* oocyte expression system [19].

1.2.3 GluN1 Subunit

GluN1 subunit identified has the amino-terminal signal sequence and four hydrophobic regions (M1–M4) near the carboxyl-terminus. GluN1 has eight splice variants of 888–941 amino acids long in their mature forms [20] (Fig. 1.1a). One alternative

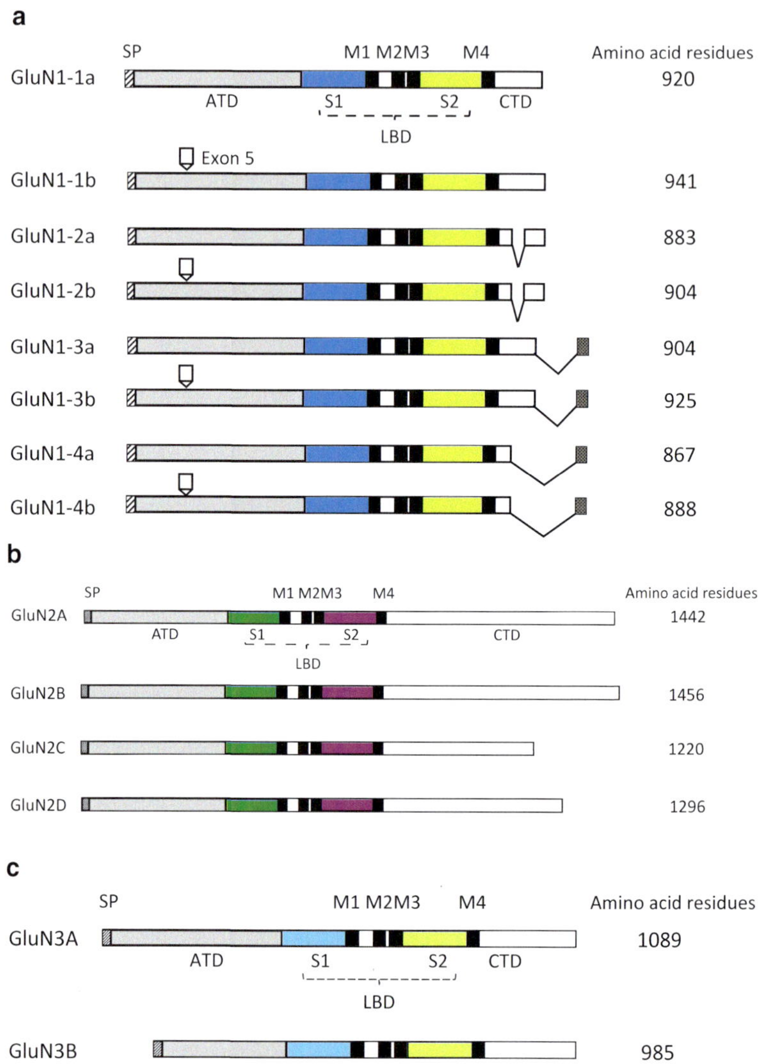

Fig. 1.1 Schematic structure of NMDAR subunits. (**a**) GluN1 and its splice variants. (**b**) GluN2 subunit. (**c**) GluN3 subunit. The putative amino-terminal signal peptide (SP) sequence, amino-terminal domain (ATD), hydrophobic regions (M1–M4), ligand-binding domain (LBD composed of S1 and S2), and carboxyl-terminal domain (CTD) are indicated

spliced exon (exon5) encods the region located in the amino-terminal domain (ATD) and contains many positively charged amino acids. Thus, this exon regulates channel activity by the binding of protons and polyamines. The carboxyl-terminal domain (CTD) of GluN1 is located in the cytoplasm and has four different amino acid sequences derived from the use of C1 and C2 or C2′ cassette exons [9, 10] (Fig. 1.1a). The different spatiotemporal expression levels and patterns of GluN1 splice variants suggest the different spatiotemporal regulatory mechanisms of splicing of GluN1 mRNA [21, 22]. The four different carboxyl-terminal regions of the GluN1 subunit derived from above-mentioned exons are involved in the functional modification, transport, surface expression, and membrane localization of the GluN1 subunit [23]. The ligand-binding domain (LBD) of GluN1 recognize D-serine or glycine, but not glutamate. The carboxyl-terminal regions are also involved in the interaction of NMDAR with many scaffold proteins and signaling molecules [24].

1.2.4 GluN2 Subunits

The GluN2 subfamily is composed of four distinct subunits, namely, GluN2A, GluN2B, GluN2C, and GluN2D. The CTDs of the GluN2 subunits are long, thus the four hydrophobic regions M1–M4 are located near the middle of the subunits. The mature form of GluN2 is 1218–1456 amino acids long (Fig. 1.1b). NMDAR is a complex including at least two GluN1 subunits and two GluN2 subunits. Thus, NMDAR is an assembly of GluN1 and GluN2 subunits assuming a dimer of heterodimer arrangements (heterotetramer) [25–27]. The LBD of GluN2 recognizes glutamate. In combination with the GluN1 subunit, each GluN2 subunit forms highly active NMDAR. The properties of NMDAR composed of GluN1 and GluN2, such as the affinity to ligands, sensitivity to antagonists and modulators, modulation of channel properties, are depend on the GluN2 subunit. The CTD of the GluN2 subunit is also involved in the interaction of the subunit with many postsynaptic scaffold proteins and signaling molecules. The GluN2A subunit has two alternative spliced carboxyl-terminals [25].

1.2.5 GluN3 Subunits

The GluN3 subfamily is composed of two distinct subunits, GluN3A and GluN3B (Fig. 1.1c). These subunits recognize glycine or D-serine. The combination of GluN1, GluN3A, and GluN3B constitutes glycine-gated cation channels in mammalian cells [28]. However, in the combination of the GluN1 and GluN2 subunits, the GluN3 subunit decreases the channel activity of NMDAR. Therefore, GluN3 is the inhibitory subunit of NMDAR [18].

1.2.6 Distribution of NMDAR Subunits

The spatiotemporal expression pattern of each NMDAR subunit suggests the functional diversity of NMDAR *in vivo*. The GluN1 subunit is expressed ubiquitously in CNS from the embryonic to adult stages [29, 30]. In embryonic mouse CNS,

GluN2B is expressed ubiquitously and GluN2D is predominantly expressed in the brainstem [29, 30]. After birth, the expression of GluN2B is restricted in the forebrain regions and the expression level of the GluN2D subunit markedly decreases. The expression of GluN2A is enhanced and ubiquitous after birth, and that of GluN2C is restricted mainly in cerebellar granule cells. The expression level of GluN3A is relatively high at younger stages and decreases later in many brain regions in rats [31]. The expression level of GluN3B is relatively high at younger stages and detected in specific regions including the pons, midbrain, medulla, and spinal cord in mice [18]. The molecular mechanism regulating these unique expression patterns of the NMDAR subunit has not been clarified. In addition to the expression of NMDAR subunits in CNS, some NMDAR subunits are detected in peripheral tissues, such as in the islets of Langerhans in the pancreas [32, 33], heart [34], bone [35], and cancer cells [36].

1.2.7 NMDAR Complex

NMDAR is composed of GluN1, GluN2, and GluN3 subunits. Synaptic NMDAR localizes to the postsynaptic density (PSD) by the interaction between its CTD and many PSD proteins to form a large NMDAR complex [24]. Furthermore, some NMDAR-associated molecules such as Neto1, EphB, and the zinc transporter Znt1 have been suggests, at least indirectly, affect NMDAR activity [37–39].

1.3 Functional Domains of NMDAR Subunits

1.3.1 Amino-Terminal Domain (ATD)

The members of the GluR channel family have 400- to 450-amino-acid-long amino-terminal domains (ATDs) (Figs. 1.1 and 1.2). The ATDs of NMDAR are not essential for channel activity, but are involved in the control of pharmacological and kinetic properties [40] and are sites of interaction with ephrin receptor [38]. The ATD of the GluN2B has a clamshell-like structure composed of two R1 and R2 domains [41]. The ATDs of GluN2A and GluN2B are the site for inhibitory Zn^{2+} binding and the GluN2B-specific ifenprodil-binding site, respectively [41].

1.3.2 Ligand-Binding Domain (LBD)

The endogenous neurotransmitter/agonist of the NMDAR is glutamate, but the full activation of NMDAR requires the concomitant binding of glutamate and a coagonist, namely, glycine or D-serine [42]. Glutamate binds to the GluN2 subunit and glycine or D-serine binds to the GluN1 and GluN3 subunits. Thus, both of the GluN1

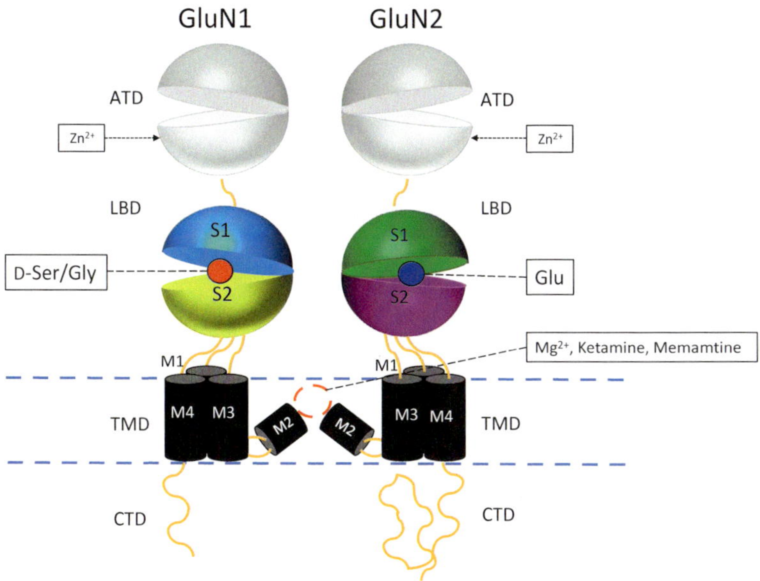

Fig. 1.2 Modular structure of heteromeric NMDAR. The amino terminal domain (ATD), ligand-binding domain (LBD) composed of S1 and S2 with agonist (Glu) and coagonist (D-Ser and Gly), transmembrane domains (TMD, composed of M1, M2, M3 and M4), and intracellular carboxyl-terminal domain (CTD) of GluN1 and GluN2 subunits are indicated. Some ligand-binding regions are shown

and GluN2 subunits are required for NMDAR to be functional. Although glycine is abundant in CNS, the coagonist-binding site of NMDAR is considered to be not saturated with glycine. Glycine in the extracelluar region is actively uptaken by the glycine transporter highly expressed in CNS [43]. Although D-serine is not involved in protein synthesis, a significant amount of D-serine has been detected in the forebrain [44]. D-Serine is produced by PLP-dependent enzyme serine racemase in CNS [45]. The spontaneous and evoked NMDAR currents are reduced after specific enzymatic degradation of D-serine [46], suggesting that D-serine is an endogenous coagonist of NMDAR. The differential roles of the two coagonists, glycine and D-serine, in the regulation of NMDAR have been extensively reviewed [47].

LBD is highly conserved in different GluR families and shares homology with bacterial periplasm amino acid binding proteins. The LBD of GluR is formed by two extracellular structures of amino acids referred to as S1 and S2 [48]. S1 locates on the extracellular amino-terminal side near M1 and S2 located on the extracellular side between M3 and M4 (Fig. 1.2). The LBD structures show a clamshell-like conformation. The recombinant fusion proteins composed of S1 and S2 from GluN1, GluN2, and GluN3 linked to the linker sequence expressed in *E. coli* are sufficient for ligand binding, and their crystal structures with ligands have been revealed [49–51]. The heterodimeric structure of the LBD complex composed of NMDAR with GluN1 and GluN2 reinforces the view that tetrameric NMDAR is a dimer of heterodimers. Over all, these domains form a shell-like structure.

1.3.3 Channel Forming Domain

NMDAR channel have unique properties compared with non-NMDAR channels. The activation of NMDAR requires the binding of two ligands (the agonist glutamate binds to GluN2 subunit and the coagonist glycine/D-serine binds to GluN1 subunit) and release from extracellular Mg^{2+} blocking in channel pore. The positive charge of Mg^{2+} is the basis of Mg^{2+} blocking of NMDAR by the electronic force to the negative charge of a neuron under the physiological membrane potential of about −70 mV. NMDAR is activated by the depolarization of membrane potential. Thus, NMDAR is a voltage-dependent and ligand-gated ion channel. This property is necessary for the detection of the coincidence of presynaptic activation (glutamate release) and postsynaptic activation (membrane depolarization). For the function of the coincidence detector, the mechanism of Mg^{2+} blocking is very important. With the opening of NMDAR ion channels, the monovalent cationic ions Na^+ and K^+ can permeate enter the channels depending on the electronic and concentration gradients. In contrast to the case of non-NMDAR channels, Ca^{2+} can enter the NMDAR channel, which induces intracellular signal transduction in the neurons. A low level of Ca^{2+} influx into neurons is necessary for the survival of neurons. However, excessive Ca^{2+} influx activates Ca^{2+}-dependent proteases and lipases, and depolarizes the mitochondrial membrane, which induce neuronal damage and death.

Agonist binding and channel gating involve three sequential steps: (1) agonist binding, (2) conformational change, such as the clam shell closure of LBD, and (3) conformational change of ion channel pore to open. Introduction of mutations in the second hydrophobic region (M2) affects the ion channel properties; thus, the second hydrophobic region is involved in the formation of channel pore. In M2, NMDAR subunits have the critical amino acid residue aspargine (N) determining Mg^{2+} blocking and Ca^{2+} permeability [52, 53]. Later analyses with the identification of the phosphorylation and glycosylation sites suggest that the topology of NMDAR subunits has three transmembrane helices (M1, M3, and M4) and one loop (M2) structure (Fig. 1.2). The M2 loop lines the inner cavity of the ion channel pore. The functionally critical N residue in the M2 of NMDAR is also involved in channel blocking by open channel blockers (PCP, ketamine, and MK-801), polyamines, and protons [54, 55].

1.3.4 Cytoplasmic Carboxyl-Terminal Domain (CTD)

The NMDAR subunit GluN1 has four distinct carboxyl-termini derived from alternative splicing and the GluN2 subunit has long carboxyl-terminal region. The CTD of NMDAR affects membrane targeting, stabilization, modification by phosphorylation and palmitoylation, and degradation [56, 57]. The CTD of NMDAR also provides interaction sites for many intracellular proteins important for signal transduction and synaptic formation, and is involved in the formation of NMDAR complexes [24].

1.3.5 Crystal Structure of Heteromeric NMDAR

The crystal structure of intact heterotetrameric NMDARs has been reported [58, 59] and the interaction sites of many ligands are mapped on the architecture of an intact NMDAR [60]. The crystal structure of NMDAR revealed the intimate association between ATDs and LBDs and the ATD-LBD interaction is fundamental to the capability of ligand binding at ATD for the propagation of conformational change and affects channel activity.

1.4 Physiological Roles of NMDAR Revealed Using Gene-Manipulated Mice

As mentioned above, NMDAR is involved in synaptic plasticity and learning and memory. Furthermore, NMDAR plays critical roles in developmental neural network formation. The physiological roles of NMDAR has been extensively examined pharmacologically using specific agonists and antagonists. In this section, I will focus on the physiological roles of each NMDAR subunit revealed by the analyses of phenotypes of subunit-specific gene knockout (KO) mice (Table 1.1). The GluN1-KO mice die immediately after birth because of respiratory failure [61, 62]. The GluN1-KO mice also show impaired formation of synapses in the somatosensory map in the brainstem. Thus, GluN1 is involved in synaptic refinement and survival of mice. The roles of GluN1 in the neuronal plasticity of the hippocampus and forebrain, higher brain functions, and neural network formation have been examined using Cre-recombinase-mediated conditional (spatiotemporal-selective) GluN1-KO mice [63–67].

The GluN2A-KO mice can grow to the adult stage and breed normally, but show impaired synaptic plasticity, spatial learning, and motor learning in the adult stage [68–70]. The GluN2B-KO mice die immediately after birth because of the impaired neural network formation and loss of the suckling response [71]. The cytoplasmic carboxyl-terminal region of GluN2B is involved in the synaptic localization and function of NMDAR, because the mice with deletion in the carboxyl-terminal region of GluN2B (GluN2B-CTKO, Table 1.1) show similar phenotypes to GluN2B-KO mice and die immediately after birth [72]. In contrast, transgenic mice overexpressing GluN2B in the forebrain show enhanced synaptic plasticity and learning abilities, and are thus called "smart mice" [73]. GluN2C-KO mice show no obvious phenotype or subtle motor dysfunctions [74–76]. GluN2D-KO mice show normal development and breeding; however, they show some indications of an impaired emotional state [77, 78].

GluN3A-KO mice show no apparent behavioral abnormalities but show subtle abnormalities in their spine morphology during development [79]. GluN3B-KO mice show moderately impaired motor learning and coordination and home cage activity [80]. These phenotypic analyses of NMDAR subunits indicate the important genetic roles of each NMDAR subunit in neural network formation, plasticity, and higher brain functions (Table 1.1).

Table 1.1 Phenotypes of NMDAR mutant mice

Mutant mice	Phenotypes [Ref. no.]
GluN1-KO	Respiratory failure and neonatal death [61]
	Impairment of synapse refinement [62]
GluN1-KO-CA1	Impairment of LTP and spatial learning [63]
	Impairment of representation of place field [64]
	Impairment of trace memory [65]
GluN1-KO-forebrain	Impairment of somatosensory neural pattern formation [66]
GluN1-KO-CA3	Impairment of associative memory recall [67]
GluN1-reduced	Schizophrenia-like behaviors [89]
GluN2A-KO	Impairment of LTP and spatial learning [68]
	Increased thresholds of LTP and contextual learning [69]
	Impairment of eye-blink trace memory [70]
	Attenuation of focal ischemic brain damage [84]
	Hyperfunction of dopaminergic and serotonergic systems [90]
GluN2B-KO	Neonatal death. Impairment of synapse refinement [71]
GluN2B-CTKO	Neonatal death. Impairment of synaptic localization of NMDAR and synapse refinement [72, 76]
GluN2B-Tg-Forebrain	Enhanced synaptic plasticity and learning [73]
GluN2C-KO	Subtle motor dysfunction [74–76]
GluN2D-KO	Impairment of emotional states [77, 78]
GluN3A-KO	Subtle abnormality in spine morphology [79]
GluN3B-KO	Moderate impairment of motor coordination [80]

CA1 hippocampal CA1-selective, *CA3* hippocampal CA3-selective, *CT* carboxy-terminal deletion, *Tg* transgenic

1.5 Pathological Roles of NMDAR

Because the hyperactivation and hypoactivation of NMDAR are involved in the many disease states, the antagonists and agonists of NMDAR have potential use for treating these disorders. In this section, I will focus on the roles of NMDAR in neurodegenerative disorders, psychiatric disorders including schizophrenia and depression, and anti-NMDAR encephalitis.

1.5.1 Neurodegenerative Disorders

NMDAR overactivation induces excessive Ca^{2+} influx and aberrant activation of many Ca^{2+}-dependent proteases and lipases [81]. This overactivation is involved in the final steps of neuronal death after stroke and in neurodegenerative disorders such as Parkinsons's disease, Alzheimer's disease (AD), Hunchington's disease, and amyotrophic lateral sclerosis [81–83]. In animal models, GluN2A-KO mice show a reduced volume of damaged areas after focal ischemia in the forebrain [84] and an oral vaccine against GluN1 provides a neuroprotective effect in an ischemic rat model [85]. In human studies, low-affinity NMDAR channel blocker memantine is used for the treatment of AD [86, 87].

1.5.2 Psychiatric Disorders

Schizophrenia is characterized by positive symptoms, negative symptoms, and cognitive dysfunction. Schizophrenia has been considered as a disease of hyperdopaminergic states. However, the drugs targeting the dopaminegic system are only effective against positive symptoms. The NMDAR hypofunction hypothesis in schizophneria is supported by the following findings, (1) NMDAR antagonists ketamine and PCP induce schizophrenia-like symptoms in humans [88], (2) Mice with reduced expression levels of the NMDAR GluN1 subunit (GluN1-reduced, Table 1.1) and GluN2A-KO mice show schizophrenia-like phenotypes [89, 90], (3) Mice with GluN1-KO selectively in parvalbumin positive GABAergic interneurons show many of the schizophrenic phenotypes [91]. These findings suggest the deficiency of NMDAR in a subclass of inhibitory interneurons underlies the pathophysiology of schizophrenia [83]. The NMDAR hypofunction hypothesis led to the idea that NMDAR potentiation may have therapeutic benefits. Treatments with NMDAR coagonists glycine [92] and D-serine [93], and with the partial agonist D-cycloserine [92] show significant efficacy in the schizophrenia patients.

The depressive state is observed in bipolar (mania and depression) and major depressive disorders. Depression has been considered as a disease with a hypomonoaminergic transmission. Antidepressants targeting the monoaminergic transmission such as selective serotonin reuptake inhibitors (SSRIs) require several weeks to exert their clinical effects [94]. In contrast, the NMDAR antagonist ketamine produces rapid (within hours) and sustained reduction in depressive symptoms in patients with treatment-resistant depression [95, 96]. Ketamine reduces the spontaneous activity of GABAergic inhibitory neurons and as a result leads to the delayed increase in the activity of excitatory neurons in rats [97]. Thus, the possible target of ketamine in depression is NMDAR in GABAergic interneurons.

1.5.3 Autoimmune Anti-NMDAR Encephalitis

The disease concept of autoimmune anti-NMDAR encephalitis with ovarian teratoma in young females has been proposed [98]. Further analyses indicate that anti-NMDAR encephalitis is characterized by memory deficits, seizures, confusion, and psychological disturbances in males and females of all ages [99]. This encephalitis is sometimes severe, potentially lethal, but treatment-responsive mediated by autoantibodies against NMDAR. Anti-NMDAR antibodies cause loss of cell surface NMDAR by antibody-mediated internalization and degradation of NMDAR [100–102]. After the detection of anti-NMDAR antibodies in the serum or CSF by ELISA and cell-based assay [103], tumor resection and immunotherapy (corticosteroids, intravenous immunoglobulins, or plasma exchange) result in a better prognosis [99]. Involvement of anti-NMDAR autoantibodies is also suggested in schizophrenia and bipolar disorder [103, 104].

1.6 Conclusions and Perspectives

After the molecular cloning of NMDAR subunits, the structural basis of important NMDAR functions such as ligand binding, channel opening, modulation, and interaction with other molecules has been revealed. Furthermore, gene manipulation of the NMDAR subunits in mice identified the physiological and pathological roles of the NMDAR subunits *in vivo*. In the future, information of the crystal structure and estimated dynamic structure of NMDAR will lead to the understanding of basis of the complex NMDAR functions and the development of novel agonists and antagonists of NMDAR for therapeutic applications.

Acknowledgements I thank Miss Kiyomi Nakagawa for the preparation of the figures. Parts of this work was supported by a Grant from the Ministry of Education, Culture, Sports, Science and Technology (MEXT) of Japan (Grant No. 25293059).

Conflict of Interest The author declare that I have no conflicts of interst with the contents of this article.

References

1. Watkins JC. The synthesis of some acidic amino acids possessing neuropharmacological activity. J Med Pharm Chem. 1962;91:1187–99.
2. Curtis DR, Watkins JC. Acidic amino acids with strong excitatory actions on mammalian neurones. J Physiol. 1963;166:1–14.
3. Bliss TVP, Collingridge GL. A synaptic model of memory: long-term potentiation in the hippocampus. Nature. 1993;361:31–9.
4. Collingridge GL, Kehl SJ, McLennan H. Excitatory amino acids in synaptic transmission in the Schaffer collateral-commissural pathway of the rat hippocampus. J Physiol. 1983;334:33–46.
5. Morris RG, Anderson E, Lynch GS, Baudry M. Selective impairment of learning and blockade of long-term potentiation by an *N*-methyl-D-aspartate receptor antagonist, AP5. Nature. 1986;319:774–6.
6. Moriyoshi K, Masu M, Ishii T, Shigemoto R, Mizuno N, Nakanishi S. Molecular cloning and characterization of the rat NMDA receptor. Nature. 1991;354:31–7.
7. Schmidt C, Hollmann M. Apparent homomeric NR1 currents observed in Xenopus oocytes are caused by an endogenous NR2 subunit. J Mol Biol. 2008;376:658–70. doi:10.1016/j.jmb.2007.11.105.

8. Yamazaki M, Mori H, Araki K, Mori KJ, Mishina M. Cloning, expression and modulation of a mouse NMDA receptor subunit. FEBS Lett. 1992;300:39–45.
9. Sugihara H, Moriyoshi K, Ishii T, Masu M, Nakanishi S. Structures and properties of seven isoforms of the NMDA receptor generated by alternative splicing. Biochem Biophys Res Commun. 1992;185:826–32.
10. Hollmann M, Boulter J, Maron C, Beasley L, Sullivan J, Pecht G, Heinemann S. Zinc potentiates agonist-induced currents at certain splice variants of the NMDA receptor. Neuron. 1993;10:943–54.
11. Meguro H, Mori H, Araki K, Kushiya E, Kutsuwada T, Yamazaki M, Kumanishi T, Arakawa M, Sakimura K, Mishina M. Functional characterization of a heteromeric NMDA receptor channel expressed from cloned cDNAs. Nature. 1992;357:70–4.
12. Kutsuwada T, Kashiwabuchi N, Mori H, Sakimura K, Kushiya E, Araki K, Meguro H, Masaki H, Kumanishi T, Arakawa M, Mishina M. Molecular diversity of the NMDA receptor channel. Nature. 1992;358:36–41.
13. Ikeda K, Nagasawa M, Mori H, Araki K, Sakimura K, Watanabe M, Inoue Y, Mishina M. Cloning and expression of the epsilon 4 subunit of the NMDA receptor channel. FEBS Lett. 1992;313:34–8.
14. Monyer H, Sprengel R, Schoepfer R, Herb A, Higuchi M, Lomeli H, Burnashev N, Sakmann B, Seeburg PH. Heteromeric NMDA receptors: molecular and functional distinction of subtypes. Science. 1992;256:1217–21.
15. Ishii T, Moriyoshi K, Sugihara H, Sakurada K, Kadotani H, Yokoi M, Akazawa C, Shigemoto R, Mizuno N, Masu M, Nakanishi S. Molecular characterization of the family of the *N*-methyl-D-aspartate receptor subunits. J Biol Chem. 1993;268:2836–43.
16. Ciabarra AM, Sullivan JM, Gahn LG, Pecht G, Heinemann S, Sevarino KA. Cloning and characterization of chi-1: a developmentally regulated member of a novel class of the ionotropic glutamate receptor family. J Neurosci. 1995;15:6498–508.
17. Sucher NJ, Akbarian S, Chi CL, Leclerc CL, Awobuluyi M, Deitcher DL, Wu MK, Yuan JP, Jones EG, Lipton SA. Developmental and regional expression pattern of a novel NMDA receptor-like subunit (NMDAR-L) in the rodent brain. J Neurosci. 1995;15:6509–20.
18. Matsuda K, Kamiya Y, Matsuda S, Yuzaki M. Cloning and characterization of a novel NMDA receptor subunit NR3B: a dominant subunit that reduces calcium permeability. Brain Res Mol Brain Res. 2002;100:43–52.
19. Chatterton JE, Awobuluyi M, Premkumar LS, Takahashi H, Talantova M, Shin Y, Cui J, Tu S, Sevarino KA, Nakanishi N, Tong G, Lipton SA, Zhang D. Excitatory glycine receptors containing the NR3 family of NMDA receptor subunits. Nature. 2002;415:793–8.
20. Mori H, Mishina M. Structure and function of the NMDA receptor channel. Neuropharmacology. 1995;34:1219–37.
21. Zhong J, Russell SL, Pritchett DB, Molinoff PB, Williams K. Expression of mRNAs encoding subunits of the *N*-methyl-D-aspartate receptor in cultured cortical neurons. Mol Pharmacol. 1994;45:846–53.
22. Laurie DJ, Seeburg PH. Regional and developmental heterogeneity in splicing of the rat brain NMDAR1 mRNA. J Neurosci. 1994;14:3180–94.
23. Lau CG, Zukin RS. NMDA receptor trafficking in synaptic plasticity and neuropsychiatric disorders. Nat Rev Neurosci. 2007;8:413–26.
24. Husi H, Ward MA, Choudhary JS, Blackstock WP, Grant SG. Proteomic analysis of NMDA receptor-adhesion protein signaling complexes. Nat Neurosci. 2000;3:661–9.
25. Traynelis SF, Wollmuth LP, McBain CJ, Menniti FS, Vance KM, Ogden KK, Hansen KB, Yuan H, Myers SJ, Dingledine R. Glutamate receptor ion channels: structure, regulation, and function. Pharmacol Rev. 2010;62:405–96. doi:10.1124/pr.109.002451.
26. Furukawa H, Singh SK, Mancusso R, Gouaux E. Subunit arrangement and function in NMDA receptors. Nature. 2005;438:185–92.
27. Salussolia CL, Prodromou ML, Borker P, Wollmuth LP. Arrangement of subunits in functional NMDA receptors. J Neurosci. 2011;31:11295–304. doi:10.1523/JNEUROSCI.5612-10.2011.

28. Smothers CT, Woodward JJ. Pharmacological characterization of glycine-activated currents in HEK 293 cells expressing *N*-methyl-D-aspartate NR1 and NR3 subunits. J Pharmacol Exp Ther. 2007;322:739–48.
29. Watanabe M, Inoue Y, Sakimura K, Mishina M. Developmental changes in distribution of NMDA receptor channel subunit mRNAs. Neuroreport. 1992;3:1138–40.
30. Watanabe M, Inoue Y, Sakimura K, Mishina M. Distinct distributions of five *N*-methyl-D-aspartate receptor channel subunit mRNAs in the forebrain. J Comp Neurol. 1993;338:377–90.
31. Wong HK, Liu XB, Matos MF, Chan SF, Pérez-Otaño I, Boysen M, Cui J, Nakanishi N, Trimmer JS, Jones EG, Lipton SA, Sucher NJ. Temporal and regional expression of NMDA receptor subunit NR3A in the mammalian brain. J Comp Neurol. 2002;450:303–17.
32. Moriyama Y, Hayashi M. Glutamate-mediated signaling in the islets of Langerhans: a thread entangled. Trends Pharmacol Sci. 2003;24:511–7.
33. Marquard J, Otter S, Welters A, Stirban A, Fischer A, Eglinger J, Herebian D, Kletke O, Klemen MS, Stožer A, Wnendt S, Piemonti L, Köhler M, Ferrer J, Thorens B, Schliess F, Rupnik MS, Heise T, Berggren PO, Klöcker N, Meissner T, Mayatepek E, Eberhard D, Kragl M, Lammert E. Characterization of pancreatic NMDA receptors as possible drug targets for diabetes treatment. Nat Med. 2015;21:363–72. doi:10.1038/nm.3822.
34. Gill SS, Pulido OM, Mueller RW, McGuire PF. Molecular and immunochemical characterization of the ionotropic glutamate receptors in the rat heart. Brain Res Bull. 1998;46:429–34.
35. Szczesniak AM, Gilbert RW, Mukhida M, Anderson GI. Mechanical loading modulates glutamate receptor subunit expression in bone. Bone. 2005;37:63–73.
36. Li L, Hanahan D. Hijacking the neuronal NMDAR signaling circuit to promote tumor growth and invasion. Cell. 2013;153:86–100. doi:10.1016/j.cell.2013.02.051.
37. Cousins SL, Innocent N, Stephenson FA. Neto1 associates with the NMDA receptor/amyloid precursor protein complex. J Neurochem. 2013;126:554–64. doi:10.1111/jnc.12280.
38. Dalva MB, Takasu MA, Lin MZ, Shamah SM, Hu L, Gale NW, Greenberg ME. EphB receptors interact with NMDA receptors and regulate excitatory synapse formation. Cell. 2000;103:945–56.
39. Mellone M, Pelucchi S, Alberti L, Genazzani AA, Di Luca M, Gardoni F. Zinc transporter-1: a novel NMDA receptor-binding protein at the postsynaptic density. J Neurochem. 2015;132:159–68. doi:10.1111/jnc.12968.
40. Yuan H, Hansen KB, Vance KM, Ogden KK, Traynelis SF. Control of NMDA receptor function by the NR2 subunit amino-terminal domain. J Neurosci. 2009;29:12045–58. doi:10.1523/JNEUROSCI.1365-09.2009.
41. Karakas E, Simorowski N, Furukawa H. Structure of the zinc-bound amino-terminal domain of the NMDA receptor NR2B subunit. EMBO J. 2009;28:3910–20. doi:10.1038/emboj.2009.338.
42. Kleckner NW, Dingledine R. Requirement for glycine in activation of NMDA-receptors expressed in Xenopus oocytes. Science. 1988;241:835–7.
43. Zafra F, Aragón C, Olivares L, Danbolt NC, Giménez C, Storm-Mathisen J. Glycine transporters are differentially expressed among CNS cells. J Neurosci. 1995;15:3952–69.
44. Hashimoto A, Nishikawa T, Hayashi T, Fujii N, Harada K, Oka T, Takahashi K. The presence of free D-serine in rat brain. FEBS Lett. 1992;296:33–6.
45. Wolosker H, Blackshaw S, Snyder SH. Serine racemase: a glial enzyme synthesizing D-serine to regulate glutamate-*N*-methyl-D-aspartate neurotransmission. Proc Natl Acad Sci U S A. 1999;96:13409–14.
46. Mothet JP, Parent AT, Wolosker H, Brady Jr RO, Linden DJ, Ferris CD, Rogawski MA, Snyder SH. D-serine is an endogenous ligand for the glycine site of the *N*-methyl-D-aspartate receptor. Proc Natl Acad Sci U S A. 2000;97:4926–31.
47. Mothet JP, Le Bail M, Billard JM. Time and space profiling of NMDA receptor co-agonist functions. J Neurochem. 2015;135:210–25. doi:10.1111/jnc.13204.

48. Stern-Bach Y, Bettler B, Hartley M, Sheppard PO, O'Hara PJ, Heinemann SF. Agonist selectivity of glutamate receptors is specified by two domains structurally related to bacterial amino acid-binding proteins. Neuron. 1994;13:1345–57.
49. Hansen KB, Tajima N, Risgaard R, Perszyk RE, Jørgensen L, Vance KM, Ogden KK, Clausen RP, Furukawa H, Traynelis SF. Structural determinants of agonist efficacy at the glutamate binding site of *N*-methyl-D-aspartate receptors. Mol Pharmacol. 2013;84:114–27. doi:10.1124/mol.113.085803.
50. Yao Y, Harrison CB, Freddolino PL, Schulten K, Mayer ML. Molecular mechanism of ligand recognition by NR3 subtype glutamate receptors. EMBO J. 2008;27:2158–70. doi:10.1038/emboj.2008.140.
51. Yao Y, Belcher J, Berger AJ, Mayer ML, Lau AY. Conformational analysis of NMDA receptor GluN1, GluN2, and GluN3 ligand-binding domains reveals subtype-specific characteristics. Structure. 2013;21:1788–99. doi:10.1016/j.str.2013.07.011.
52. Mori H, Masaki H, Yamakura T, Mishina M. Identification by mutagenesis of a Mg^{2+}-block site of the NMDA receptor channel. Nature. 1992;358:673–5.
53. Burnashev N, Schoepfer R, Monyer H, Ruppersberg JP, Günther W, Seeburg PH, Sakmann B. Control by asparagine residues of calcium permeability and magnesium blockade in the NMDA receptor. Science. 1992;257:1415–9.
54. Yamakura T, Mori H, Masaki H, Shimoji K, Mishina M. Different sensitivities of NMDA receptor channel subtypes to non-competitive antagonists. Neuroreport. 1993;4:687–90.
55. Kashiwagi K, Masuko T, Nguyen CD, Kuno T, Tanaka I, Igarashi K, Williams K. Channel blockers acting at *N*-methyl-D-aspartate receptors: differential effects of mutations in the vestibule and ion channel pore. Mol Pharmacol. 2002;61:533–45.
56. Vissel B, Krupp JJ, Heinemann SF, Westbrook GL. A use-dependent tyrosine dephosphorylation of NMDA receptors is independent of ion flux. Nat Neurosci. 2001;4:587–96.
57. Hayashi T, Thomas GM, Huganir RL. Dual palmitoylation of NR2 subunits regulates NMDA receptor trafficking. Neuron. 2009;64:213–26. doi:10.1016/j.neuron.2009.08.017.
58. Karakas E, Furukawa H. Crystal structure of a heterotetrameric NMDA receptor ion channel. Science. 2014;344:992–7. doi:10.1126/science.1251915.
59. Lee CH, Lü W, Michel JC, Goehring A, Du J, Song X, Gouaux E. NMDA receptor structures reveal subunit arrangement and pore architecture. Nature. 2014;511:191–7. doi:10.1038/nature13548.
60. Regan MC, Romero-Hernandez A, Furukawa H. A structural biology perspective on NMDA receptor pharmacology and function. Curr Opin Struct Biol. 2015;33:68–75. doi:10.1016/j.sbi.2015.07.012.
61. Forrest D, Yuzaki M, Soares HD, Ng L, Luk DC, Sheng M, Stewart CL, Morgan JI, Connor JA, Curran T. Targeted disruption of NMDA receptor 1 gene abolishes NMDA response and results in neonatal death. Neuron. 1994;13:325–38.
62. Li Y, Erzurumlu RS, Chen C, Jhaveri S, Tonegawa S. Whisker-related neuronal patterns fail to develop in the trigeminal brainstem nuclei of NMDAR1 knockout mice. Cell. 1994;76:427–37.
63. Tsien JZ, Huerta PT, Tonegawa S. The essential role of hippocampal CA1 NMDA receptor-dependent synaptic plasticity in spatial memory. Cell. 1996;87:1327–38.
64. McHugh TJ, Blum KI, Tsien JZ, Tonegawa S, Wilson MA. Impaired hippocampal representation of space in CA1-specific NMDAR1 knockout mice. Cell. 1996;87:1339–49.
65. Huerta PT, Sun LD, Wilson MA, Tonegawa S. Formation of temporal memory requires NMDA receptors within CA1 pyramidal neurons. Neuron. 2000;25:473–80.
66. Iwasato T, Datwani A, Wolf AM, Nishiyama H, Taguchi Y, Tonegawa S, Knöpfel T, Erzurumlu RS, Itohara S. Cortex-restricted disruption of NMDAR1 impairs neuronal patterns in the barrel cortex. Nature. 2000;406:726–31.
67. Nakazawa K, Quirk MC, Chitwood RA, Watanabe M, Yeckel MF, Sun LD, Kato A, Carr CA, Johnston D, Wilson MA, Tonegawa S. Requirement for hippocampal CA3 NMDA receptors in associative memory recall. Science. 2002;297:211–8.

68. Sakimura K, Kutsuwada T, Ito I, Manabe T, Takayama C, Kushiya E, Yagi T, Aizawa S, Inoue Y, Sugiyama H, Mishina M. Reduced hippocampal LTP and spatial learning in mice lacking NMDA receptor epsilon 1 subunit. Nature. 1995;373:151–5.
69. Kiyama Y, Manabe T, Sakimura K, Kawakami F, Mori H, Mishina M. Increased thresholds for long-term potentiation and contextual learning in mice lacking the NMDA-type glutamate receptor epsilon1 subunit. J Neurosci. 1998;18:6704–12.
70. Kishimoto Y, Kawahara S, Mori H, Mishina M, Kirino Y. Long-trace interval eyeblink conditioning is impaired in mutant mice lacking the NMDA receptor subunit epsilon 1. Eur J Neurosci. 2001;13:1221–7.
71. Kutsuwada T, Sakimura K, Manabe T, Takayama C, Katakura N, Kushiya E, Natsume R, Watanabe M, Inoue Y, Yagi T, Aizawa S, Arakawa M, Takahashi T, Nakamura Y, Mori H, Mishina M. Impairment of suckling response, trigeminal neuronal pattern formation, and hippocampal LTD in NMDA receptor epsilon 2 subunit mutant mice. Neuron. 1996;16:333–44.
72. Mori H, Manabe T, Watanabe M, Satoh Y, Suzuki N, Toki S, Nakamura K, Yagi T, Kushiya E, Takahashi T, Inoue Y, Sakimura K, Mishina M. Role of the carboxy-terminal region of the GluR epsilon2 subunit in synaptic localization of the NMDA receptor channel. Neuron. 1998;21:571–80.
73. Tang YP, Shimizu E, Dube GR, Rampon C, Kerchner GA, Zhuo M, Liu G, Tsien JZ. Genetic enhancement of learning and memory in mice. Nature. 1999;401:63–9.
74. Ebralidze AK, Rossi DJ, Tonegawa S, Slater NT. Modification of NMDA receptor channels and synaptic transmission by targeted disruption of the NR2C gene. J Neurosci. 1996;16:5014–25.
75. Kadotani H, Hirano T, Masugi M, Nakamura K, Nakao K, Katsuki M, Nakanishi S. Motor discoordination results from combined gene disruption of the NMDA receptor NR2A and NR2C subunits, but not from single disruption of the NR2A or NR2C subunit. J Neurosci. 1996;16:7859–67.
76. Sprengel R, Suchanek B, Amico C, Brusa R, Burnashev N, Rozov A, Hvalby O, Jensen V, Paulsen O, Andersen P, Kim JJ, Thompson RF, Sun W, Webster LC, Grant SG, Eilers J, Konnerth A, Li J, McNamara JO, Seeburg PH. Importance of the intracellular domain of NR2 subunits for NMDA receptor function in vivo. Cell. 1998;92:279–89.
77. Ikeda K, Araki K, Takayama C, Inoue Y, Yagi T, Aizawa S, Mishina M. Reduced spontaneous activity of mice defective in the epsilon 4 subunit of the NMDA receptor channel. Brain Res Mol Brain Res. 1995;33:61–71.
78. Miyamoto Y, Yamada K, Noda Y, Mori H, Mishina M, Nabeshima T. Lower sensitivity to stress and altered monoaminergic neuronal function in mice lacking the NMDA receptor epsilon 4 subunit. J Neurosci. 2002;22:2335–42.
79. Das S, Sasaki YF, Rothe T, Premkumar LS, Takasu M, Crandall JE, Dikkes P, Conner DA, Rayudu PV, Cheung W, Chen HS, Lipton SA, Nakanishi N. Increased NMDA current and spine density in mice lacking the NMDA receptor subunit NR3A. Nature. 1998;393:377–81.
80. Niemann S, Kanki H, Fukui Y, Takao K, Fukaya M, Hynynen MN, Churchill MJ, Shefner JM, Bronson RT, Brown Jr RH, Watanabe M, Miyakawa T, Itohara S, Hayashi Y. Genetic ablation of NMDA receptor subunit NR3B in mouse reveals motoneuronal and nonmotoneuronal phenotypes. Eur J Neurosci. 2007;26:1407–20.
81. Choi DW. Glutamate neurotoxicity and diseases of the nervous system. Neuron. 1988;1:623–34.
82. Olney JW. Excitotoxin-mediated neuron death in youth and old age. Prog Brain Res. 1990;86:37–51.
83. Zhou Q, Sheng M. NMDA receptors in nervous system diseases. Neuropharmacology. 2013;74:69–75. doi:10.1016/j.neuropharm.2013.03.030.
84. Morikawa E, Mori H, Kiyama Y, Mishina M, Asano T, Kirino T. Attenuation of focal ischemic brain injury in mice deficient in the epsilon1 (NR2A) subunit of NMDA receptor. J Neurosci. 1998;18:9727–32.

85. During MJ, Symes CW, Lawlor PA, Lin J, Dunning J, Fitzsimons HL, Poulsen D, Leone P, Xu R, Dicker BL, Lipski J, Young D. An oral vaccine against NMDAR1 with efficacy in experimental stroke and epilepsy. Science. 2000;287:1453–60.
86. Lipton SA. Failures and successes of NMDA receptor antagonists: molecular basis for the use of open-channel blockers like memantine in the treatment of acute and chronic neurologic insults. NeuroRx. 2004;1:101–10.
87. Mota SI, Ferreira IL, Rego AC. Dysfunctional synapse in Alzheimer's disease—a focus on NMDA receptors. Neuropharmacology. 2014;76:16–26. doi:10.1016/j.neuropharm.2013.08.013.
88. Coyle JT. Glutamate and schizophrenia: beyond the dopamine hypothesis. Cell Mol Neurobiol. 2006;26:365–84.
89. Mohn AR, Gainetdinov RR, Caron MG, Koller BH. Mice with reduced NMDA receptor expression display behaviors related to schizophrenia. Cell. 1999;98:427–36.
90. Miyamoto Y, Yamada K, Noda Y, Mori H, Mishina M, Nabeshima T. Hyperfunction of dopaminergic and serotonergic neuronal systems in mice lacking the NMDA receptor epsilon1 subunit. J Neurosci. 2001;21:750–7.
91. Belforte JE, Zsiros V, Sklar ER, Jiang Z, Yu G, Li Y, Quinlan EM, Nakazawa K. Postnatal NMDA receptor ablation in corticolimbic interneurons confers schizophrenia-like phenotypes. Nat Neurosci. 2010;13:76–83. doi:10.1038/nn.2447.
92. Heresco-Levy U, Javitt DC. Comparative effects of glycine and D-cycloserine on persistent negative symptoms in schizophrenia: a retrospective analysis. Schizophr Res. 2004;66:89–96.
93. Tsai GE, Lin PY. Strategies to enhance *N*-methyl-D-aspartate receptor-mediated neurotransmission in schizophrenia, a critical review and meta-analysis. Curr Pharm Des. 2010;16:522–37.
94. Rush AJ, Trivedi MH, Wisniewski SR, Nierenberg AA, Stewart JW, Warden D, Niederehe G, Thase ME, Lavori PW, Lebowitz BD, McGrath PJ, Rosenbaum JF, Sackeim HA, Kupfer DJ, Luther J, Fava M. Acute and longer-term outcomes in depressed outpatients requiring one or several treatment steps: a STAR*D report. Am J Psychiatry. 2006;163:1905–17.
95. Berman RM, Cappiello A, Anand A, Oren DA, Heninger GR, Charney DS, Krystal JH. Antidepressant effects of ketamine in depressed patients. Biol Psychiatry. 2000;47:351–4.
96. Zarate Jr CA, Singh JB, Carlson PJ, Brutsche NE, Ameli R, Luckenbaugh DA, Charney DS, Manji HK. A randomized trial of an N-methyl-D-aspartate antagonist in treatment-resistant major depression. Arch Gen Psychiatry. 2006;63:856–64.
97. Homayoun H, Moghaddam B. NMDA receptor hypofunction produces opposite effects on prefrontal cortex interneurons and pyramidal neurons. J Neurosci. 2007;27:11496–500.
98. Dalmau J, Gleichman AJ, Hughes EG, Rossi JE, Peng X, Lai M, Dessain SK, Rosenfeld MR, Balice-Gordon R, Lynch DR. Anti-NMDA-receptor encephalitis: case series and analysis of the effects of antibodies. Lancet Neurol. 2008;7:1091–8. doi:10.1016/S1474-4422(08)70224-2.
99. Titulaer MJ, McCracken L, Gabilondo I, Armangué T, Glaser C, Iizuka T, Honig LS, Benseler SM, Kawachi I, Martinez-Hernandez E, Aguilar E, Gresa-Arribas N, Ryan-Florance N, Torrents A, Saiz A, Rosenfeld MR, Balice-Gordon R, Graus F, Dalmau J. Treatment and prognostic factors for long-term outcome in patients with anti-NMDA receptor encephalitis: an observational cohort study. Lancet Neurol. 2013;12:157–65. doi:10.1016/S1474-4422(12)70310-1.
100. Dalmau J, Lancaster E, Martinez-Hernandez E, Rosenfeld MR, Balice-Gordon R. Clinical experience and laboratory investigations in patients with anti-NMDAR encephalitis. Lancet Neurol. 2011;10:63–74. doi:10.1016/S1474-4422(10)70253-2.
101. Hughes EG, Peng X, Gleichman AJ, Lai M, Zhou L, Tsou R, Parsons TD, Lynch DR, Dalmau J, Balice-Gordon RJ. Cellular and synaptic mechanisms of anti-NMDA receptor encephalitis. J Neurosci. 2010;30:5866–75. doi:10.1523/JNEUROSCI.0167-10.2010.
102. Takano S, Takahashi Y, Kishi H, Taguchi Y, Takashima S, Tanaka K, Muraguchi A, Mori H. Detection of autoantibody against extracellular epitopes of *N*-methyl-D-aspartate receptor by cell-based assay. Neurosci Res. 2011;71:294–302. doi:10.1016/j.neures.2011.07.1834.

103. Sinmaz N, Amatoury M, Merheb V, Ramanathan S, Dale RC, Brilot F. Autoantibodies in movement and psychiatric disorders: updated concepts in detection methods, pathogenicity, and CNS entry. Ann N Y Acad Sci. 2015;1351:22–38. doi:10.1111/nyas.12764.
104. León-Caballero J, Pacchiarotti I, Murru A, Valentí M, Colom F, Benach B, Pérez V, Dalmau J, Vieta E. Bipolar disorder and antibodies against the N-methyl-d-aspartate receptor: a gate to the involvement of autoimmunity in the pathophysiology of bipolar illness. Neurosci Biobehav Rev. 2015;55:403–12. doi:10.1016/j.neubiorev.2015.05.012.

Chapter 2
Synaptic and Extra-Synaptic NMDA Receptors in the CNS

Thomas Papouin and Stéphane H.R. Oliet

Abstract The *N*-methyl D-aspartate receptor (NMDAR) is a ligand-gated ion channel that binds the neurotransmitter glutamate. It was pharmacologically identified and differentiated from other ionotropic amino-acid receptors at excitatory synapses in the late 70s for it is activated by NMDA and not kainate. Due to its large calcium conductance, it is involved in many physiological and pathological phenomena, the most notorious of which is synaptic plasticity, considered to be the molecular substrate of learning and memory. During the 40 years that followed their discovery, and owing to other unique properties such as their magnesium-block that makes them key "coincidence detectors", NMDARs have been mostly studied at synapses. Yet, NMDARs exhibit a great number of other fundamental features that have remained unknown, underappreciated or challenging to study, and that have only become the focus of intense investigation over the past decade. These properties, such as the co-agonist-gating or the subcellular compartmentalization, greatly contribute to the functional diversity of NMDARs and will be the focus of this chapter as they are greatly relevant in the context of their physiological and pathological impact on the central nervous system.

Keywords NMDA receptor • Glutamate receptor • Co-agonist • D-Serine • Glycine • Subcellular localization • Extrasynaptic • Excitotoxicity • Synaptic plasticity • Subunit composition • D-Aspartate • Slow inward currents • Tonic current • Astrocytes • Glia

With over 1500 publications each year since the late 1990s', the *N*-methyl D-aspartate receptor (NMDAR) is the most investigated receptor in the field of neurosciences. Nearly 40 years after their pharmacological identification [1, 2], an overwhelming wealth of data has become available about these glutamate-gated

T. Papouin
Department of Neuroscience, Tufts University School of Medicine, Boston, MA, USA

S.H.R. Oliet (✉)
Neurocentre Magendie, Inserm U1215 and Université de Bordeaux, Bordeaux, France
e-mail: stephane.oliet@inserm.fr

K. Hashimoto (ed.), *The NMDA Receptors*, The Receptors 30,
DOI 10.1007/978-3-319-49795-2_2

ionotropic receptors, from their fine crystallographic structures and electrophysiological properties, to the molecular basis of their functional diversity, their central role in the mechanisms of synaptic plasticity, learning, memory and their direct implication in a variety of diseases.

In this chapter, we will focus on the diversity of NMDARs subtypes and functions in the central nervous system (CNS) in light of the most recent advances in the field. We will focus on a newly discovered source of NMDARs diversity that, similar to their subunit composition, is thought to impact their function: the subcellular location. Indeed, NMDARs are mostly studied at synapses, where glutamate is released, but they also exist at pre-synaptic and extra-synaptic sites where they seem to be engaged in different processes.

As "famous" as they might be, some aspects of these receptors still suffer from a paucity of information. While this is obviously true for most of the properties of extra-synaptic NMDARs which are notoriously hard to study, this is surprisingly true as well for key features of synaptic NMDARs such as their co-agonist gating. Where data are lacking, we will dash this chapter with provocative thoughts that might help shedding a new light on some aspects of NMDARs and offer new perspectives about their roles in the central nervous system (CNS).

NMDARs are distinctive from other glutamate receptors by their remarkable molecular diversity. For the purpose of this introduction, we shall just say that NMDARs assemble as heterotetramers made of two GluN1-subunits combined with two other subunits of the GluN2 (A–D) or GluN3 (A and B) family. Each of these possible combinations provides particular properties to the receptor subtype they make up [3, 4], including fundamental aspects such as the magnesium block, the affinity for agonists and the Ca^{2+} permeability of the ion pore. Combined with the spatial (brain region-specific) and temporal (developmental) profile of NMDAR subunits expression in the CNS, this results in a very complex and heterogeneous picture of NMDAR properties in a given area of the CNS at a given time of development or adulthood. This is rendered even more complex by the fact that a given NMDAR subtype can be found at diverse subcellular locations at the surface of a single neuron, where it is believed to play distinct roles.

This probably explains why the use of subunit-specific NMDARs antagonists has proven disappointing in a clinical context, considering the plethora of fundamental functions involving NMDARs in the CNS on one hand, and the different, sometimes opposing, roles that a subtype of NMDAR can endorse depending on its brain or cellular location. Indeed, the discovery that many of the properties of NMDARs are highly sensitive to the subunit composition of the receptor generated a great drive to attribute specific functions to particular subtypes of NMDAR and to develop subunit-specific pharmacological agonists, antagonists and modulators, aiming at treating major neuropathological and neurodegenerative diseases (from ischemia to schizophrenia, see following chapters). Besides this great potential and besides the wealth of data available on NMDARs, most of these therapeutic compounds have failed in clinical trial due to adverse side effects and/or lack of efficacy [5]. This type of observations has contributed to make it clear that the cellular and subcellular location, combined with NMDAR subunit composition, could be one of the most important features dictating the functions in which a given subtype of NMDAR is involved. In this chapter, we will focus on neuronal NMDARs, but NMDARs can also be found on blood vessels [6, 7], astrocytes [8] and oligodendrocytes [9].

2.1 Synaptic Versus Extra-Synaptic NMDARs: Location, a Source of Functional Diversity

2.1.1 What Is Synaptic and What Is Not?

While the vision we often have of synapses is very 'graphic' owing to their specialized morphology, the definition of the synaptic space remains highly functional. Along those lines, post-synaptic NMDARs are conventionally considered as 'synaptic' if they are recruited during low frequency presynaptic activity, which includes low frequency stimulation of axon terminals and miniature/spontaneous vesicular release events [10–14]. Such a definition has the advantage of being very intuitive: An NMDAR is synaptic if it is activated by synaptic activity. However, it provides no insight about the *actual* localization of receptors participating to synaptic transmission, in particular because the geometry of the cleft and adjacent extra-synaptic space, the rate of glutamate uptake, the presynaptic site of vesicular release, the concentration of transmitter in vesicles, and many other parameters greatly influence the probability that a receptor anchored at a given location will bind glutamate [15].

From the morphological point of view, a receptor is often considered synaptic if it lays no more than 100 nm from the edge of the post-synaptic density (PSD) [16, 17]. This often implies that the receptor faces the presynaptic terminal. In a similar manner that the functional definition is satisfying at the electrophysiological level, this anatomical definition fulfills our need to *visualize* a synaptic receptor *at* the synapse. Unfortunately, it is very unlikely that these two definitions encompass the same synaptic space and the same pool of receptors, and the delineation of the synaptic vs. extra-synaptic space is probably a case by case matter given the diversity of synaptic features throughout the different regions of the CNS. Meanwhile, we are still lacking a clear definition of the boundaries of the synaptic space that would best account for both functional and morphological considerations and this would certainly need to be addressed.

2.1.2 How Are NMDARs Organized at Synapses?

The precise organization of NMDARs within the synapses is not yet fully resolved and seems to be specific for each type of synapse. For instance, the density of NMDARs was observed to peak at the edge, or past the edge, of PSD at ganglion cell synapses in the retina. This seems particularly true for GluN2B-containing receptors [18]. However, this finding does not seem to be the case at other synapses, such as in the hippocampus, and this could be due to the nature and distribution of the post-synaptic intracellular partners present at each synapse (see Sect. 2.1.4). NMDARs also exist at pre-synaptic locations [17, 19, 20]. Though they are, strictly speaking, extra-synaptic given their distance to the PSD and their location on a different cell, they are often considered synaptic because they are thought to contribute to synaptic function. Whether pre-synaptic NMDARs are controlled by synaptic glutamate spill-over, by surrounding glial processes or by both, is not clear. Examples of either situation have been documented [19, 20] and it seems that the function of these receptors mostly consists in regulating the release probability of the pre-synaptic element, as could be expected given the Ca^{2+} permeability of NMDARs.

2.1.3 How Are NMDARs Organized at Extra-Synaptic Sites?

NMDARs were discovered at non-synaptic locations in the mid-1990s [21–31] and they were initially thought to be very similar to synaptic NMDARs, because the channel behavior in excised patches (extra-synaptic NMDARs) was directly related to the macroscopic properties of the excitatory post-synaptic current (EPSC) recorded from synapses (synaptic NMDARs) [23]. Subsequent functional comparison of synaptic NMDAR-mediated currents and outside-out patches, however, has since revealed that "different NMDAR subtypes are expressed in sub-synaptic and extra-synaptic compartments" [23, 25, 27]. For 20 years, our approach of extra-synaptic NMDARs as remained bound to this idea [25] and the first study actually dedicated to the organization of NMDARs at extra-synaptic sites, rather than to their subunit composition, was only published in 2010 [17]. It revealed that extra-synaptic NMDARs are found in cultures as well as on brain sections, both at early developmental stage and in adulthood. This was concordant with electrophysiological evidence that extra-synaptic NMDARs represent an estimated 1/3 of the total NMDAR population on hippocampal neurons in young adult rats [12] and as high as ¾ of all NMDARs in the immature hippocampus [14]. At extra-synaptic sites, NMDARs form clusters that are not evenly distributed on dendrites. Instead, they are preferentially associated with specific portions of the dendritic shaft that are contacted by other cells such as by glial cell processes (~30 %), axon-like processes (~50 %) or other dendrites (~20 %). This is in remarkable agreement with earlier observations [24] that found extra-synaptic GluN1 immuno-gold labeling consistently localized between dendrites and astrocytic processes in adult somatosensory cortex. This very specific localization supports a role for NMDARs in reciprocal neuron–glia communication (Fig. 2.1). At synapses, the PSD, a dense and intricate ensemble of intracellular partners held together with scaffolding proteins that play a key role in regulating glutamate receptors signaling and synaptic architecture, typically spreads over 195 nm on average in the area CA1 of the hippocampus [32] or over 260 nm at the particular CA3-CA1 synapse [33]. Intriguingly, extra-synaptic

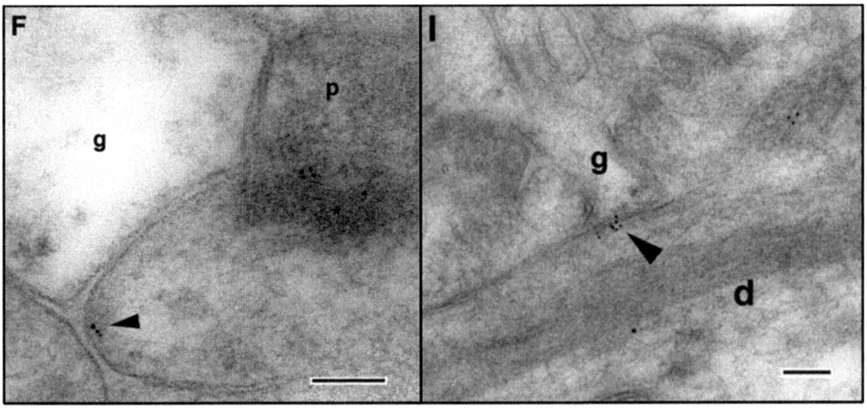

Fig. 2.1 Immuno-gold labelling of GluN1 antibody in the area CA1 of adult hippocampus. Note the presence of GluN1 labeling at remote distance from the pre-synaptic terminal (p) but in close proximity with a glial (g) process (*arrowhead*, *panel F*). On *panel I*, GluN1 subunits are clearly clustered at the surface of a dendrite (d) where a glial process makes a "synapse-like" contact (*arrowhead*). *Scale bar*: 100 nm. Adapted from [17] Figure 5

NMDARs were found to gather over an area of less than 100 nm width on average with no clear PSD-like sub-membrane electron-dense structure. This could suggest that the sites where extra-synaptic NMDARs accumulate are comprised of no (or few) other types of receptors, thus requiring limited sub-membrane partners and anchoring proteins, and are dedicated to NMDAR-mediated extra-synaptic signaling only. This would explain why they are found in a narrower region, and why NMDAR-containing extra-synaptic sites lack a clear PSD-like electron density. None the less, much like their synaptic homologues, clusters of NMDARs at extra-synaptic sites seem to associate with intercellular adhesion molecules such as β-catenins and with the scaffolding proteins SAP-102 (synapse associated protein-102) and PSD95 (post-synaptic density protein 95). Therefore, it seems that the molecular organization of synaptic and extra-synaptic NMDARs at the plasma membrane follow the same elementary rules, even though the molecular details of the intracellular partners involved in anchoring NMDARs at extra-synaptic sites remain to be elucidated.

Taken altogether, these evidence about their organization strongly suggest that extra-synaptic NMDARs represent a functionally distinct pool of receptors, specifically tethered to this location, and not a pool of rogue receptors escaped from synaptic trapping or a mere reserve pool of receptors waiting to be recruited to synapses. Instead, these observations could be taken as evidence that extra-synaptic receptors, much like their synaptic counterpart, are engaged in cell-to-cell signaling. The finding that these receptors are clustered to specific cell contact areas, such as specialized neuron–glia appositions, indeed suggests that they may be engaged in a separate signaling function independent from their synaptic homologues. This view is also supported by the functional demonstration that synaptic and extra-synaptic

NMDARs form distinct and stable pools of receptors at the surface of CA1 neurons [12]. These data also challenge our primitive vision of extra-synaptic NMDARs as an ensemble of highly mobile and randomly distributed receptors as was suggested by work in cultures [34–36].

2.1.4 NMDARs Are Mobile

All of these views, and most of the work performed on NMDARs, are somewhat based on the assumption that NMDARs are mostly static over short periods of time, in particular at the time-scale of synaptic transmission. Surprisingly, this may not be true. Indeed, recent work carried out in cultures has demonstrated the high motility of glutamate receptors. Although initially reported decades ago [37, 38], membrane diffusion of neurotransmitter receptors only recently emerged as a cellular pathway involved in the regulation of synaptic receptor content and distribution [39], perhaps because this has been studied more accurately with the development of techniques such as fluorescence recovery after photo-bleaching (FRAP) microscopy and *quantum* dots (Q-dots) over the last years. This so-called *lateral diffusion* of NMDARs (i.e. rapid and seemingly Brownian movement of the receptor at the surface of the plasma membrane) was elegantly evidenced at the surface of young hippocampal neurons by Tovar and Westbrook in 2002 [36]. Using the NMDAR antagonist MK-801, an activity-dependent and irreversible open-channel blocker, to completely block synaptic NMDARs and abolish synaptic NMDAR-mediated currents, they were able to show a partial recovery of synaptic NMDAR-mediated responses within minutes, demonstrating that synapses had been refilled with unblocked NMDARs from extra-synaptic sites via lateral diffusion of receptors and that blocked synaptic receptors had diffused away from the PSD to the extra-synaptic compartment.

Today, we can directly image the trajectory of single-NMDARs in real-time with the use of Q-dotes (Fig. 2.2) and we know that approximately 30–40 % of surface NMDARs are mobile with an average diffusion coefficient of 0.05 $\mu m^2/s$. In comparison, 50 % of surface AMPARs are mobile in basal conditions with an instantaneous diffusion coefficient in the range of 100–1000 $\mu m^2/s$. It was shown that about a third of NMDARs traffic between the synaptic and extra-synaptic compartment, which means that NMDARs reside at the synapse for a matter of minutes, rather than days as previously thought [35, 40, 41]. The lateral diffusion of NMDARs is strongly influenced by the subunit composition of the receptor. In particular, GluN2B-containing NMDARs appear more mobile than GluN2A-containing NMDARs, in line with predictions that could be made from their interactions with distinct intracellular partners. Whether this mobility is affected by activity, in particular by the binding of glutamate, remains unclear. It was shown *in vivo* that the blockade of NMDARs leads to a rapid and striking redistribution of GluN2B subunits away from synapses [30]. On the other hand, inhibiting NMDARs does not affect diffusion properties in brain slices [13].

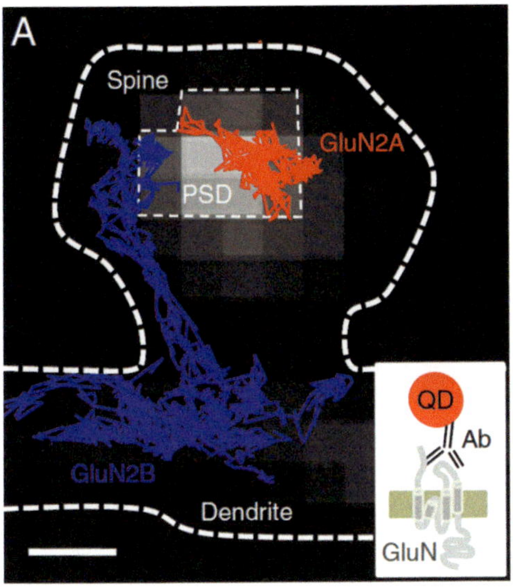

Fig. 2.2 Illustration of the lateral diffusion of GluN2B- and GluN2A-countaining NMDARs at the surface of a dendritic spine of a cultured hippocampal neuron. The representative trajectory (over 50 ms) of a surface GluN2A-containing NMDAR is depicted in *red*, that of a GluN2B-containing NMDAR in *blue*. They were imaged using single *Quantum*-Dot (QD) tracking approach (*inset, lower right*). Each trajectory represents the diffusion of a single particle-receptor complex. The GluN2A–QD is confined in the PSD at the spine head while the GluN2B–QD diffuses from the head of the spine to the dendritic shaft. Scale bar = 200 nm. Adapted from [35]

2.2 NMDAR Intracellular Partners at Synapses and Extra-Synaptic Sites

Membrane-associated guanylate kinases (MAGUKs), such as PSD-95 and SAP-102, are direct intracellular partners that coordinate trafficking, anchoring, and signaling of NMDARs (among other receptors) by interacting with their cytoplasmic tail and by recruiting other signaling and scaffolding proteins at the PSD. It is still unclear to what extent particular NMDAR subtypes associate preferentially with specific MAGUKs but it is generally accepted that mature synapses are enriched in PSD-95, while immature/developing synapses contain SAP-102 [42–44]. Consequently, it is thought that NMDARs are anchored at mature synapses primarily via interactions with PSD-95 [45], while interactions with SAP-102 is responsible for their trafficking and anchoring at immature synapses. SAP-102 is also expressed in the adult but a strong competition with PSD-95 for insertion into the PSD causes its exclusion from the synapse. SAP-102 is thus thought to be enriched at peri-synaptic or extra-synaptic locations in adult. Importantly, it can still be found at synapses [42, 44], such that this view is only a gross approximation. According to this view, in the adult, interaction of NMDARs with SAP-102 would result in trafficking and addressing of the receptor to an extra-synaptic location. Because the spatial expression of SAP-102 seems less compartmentalized than that of PSD-95 this would also result in a higher mobility of the receptor. New data also support a role for SAP-102 in the synaptic clearance of NMDARs, indicating that it could play a role in regulating receptor content at synapses over time, which is very interesting in the context of activity-dependent reorganization of NMDARs that was observed at many occasions [30]. At extra-synaptic sites, it is unclear what MAGUK prevails, but consistent with the description above, the SAP-102/PSD-95 ratio is overall enriched in favor of a higher content of SAP-102. Petralia et al. found that (in cultures) SAP-102 is not enriched at synapses but rather expressed evenly in both extra-synaptic and synaptic compartments, in contrast with PSD-95 which they found to be five times more concentrated at synapses. The relative functional enrichment of SAP-102 at extra-synaptic sites would thus result solely from the preferential location of PSD-95 at synapses. Whether the nature of intracellular partners present at extra-synaptic locations varies according to the identity of the cellular processes that NMDARs are facing is unknown, but is an interesting possibility.

2.3 Subunit Composition at Synaptic and Extrasynaptic Sites

Details about the assembling of NMDARs subunits into a functional tetrameric receptor can be found in any of the many reviews from Dr. Paoletti. This section will focus on the functional properties that subunit composition confers to the receptor. Briefly, the assembling of NMDAR requires four subunits, two of which are necessarily the obligatory GluN1 subunit (any of its two splice variants). The two remaining subunits can be any of the 4 GluN2 subunits (A–D) and/or of the 2 GluN3 subunits (A and B). The most notorious of these combinations are the GluN2A-NMDARs (di-heteromers made of 2 GluN1 assembled with 2 GluN2B) and the GluN2B-NMDARs (GluN1-GluN2B di-heteromers) for which we have had very selective and efficient antagonists since the late 1980s. Like we already mentioned, the identity of the subunits that assemble confer a unique set of properties to the NMDAR they form together (Table 2.1). Owing to differences in their intercellular C-terminal domain (CTD), GluN2 subunits also interact with different scaffolding proteins, which strongly influences the surface localization and diffusion of NMDAR subtypes [3, 4], and this is thought to be the basis for location-dependent differences in NMDAR subunit composition. For instance, it is generally believed that GluN2A-NMDARs associate preferentially with PSD-95, and this interaction is responsible for trapping these receptors at synapses [35, 42–44, 46, 47]. GluN2B-NMDARs on the other hand tend to interact with SAP-102 [42–44]. According to the view described in Sect. 2.2 this would make GluN2B-NMDARs more mobile and addressed at extra-synaptic sites, which is remarkably consistent with existing data (see Sect. 2.1.2 and Fig. 2.2 above, and Sect. 2.3.2 below).

Table 2.1 Effect of different GluN1-GluN2 (and GluN3A) di-heteromeric subunits assembling on some of the basic properties of NMDAR such as the affinity for agonists glutamate and glycine, affinity of the magnesium block, sensitivity to pH, open probability and proven specific non-competitive antagonists

Di-heteromeric receptor	GlutamateEC_{50} (μM)	GlycineEC_{50}	Mg^{2+} IC_{50} (at −100 mV) (μM)	pH_{IC50}	Open probability (Po)	Non-competitive antagonists
GluN2A-NMDAR	4	1.5 μM	12	6.9	0.5	Zinc, in Tricine
GluN2B-NMDAR	2	0.4 μM	12	7.5	0.1	Ifenprodil, Ro25-6981
GluN2C-NMDAR	1	0.3 μM	2	6.6	0.01	–
GluN2D-NMDAR	0.4	0.1 μM	2	7.5	0.01	–
GluN3A-NMDAR	–	40–500 nM	> 100	–	–	–

2.3.1 Subunit Composition of Synaptic NMDARs Is Highly Variable in Time and Space

NMDAR subunit composition changes over time throughout the CNS, as evidenced by their mRNA profile characterized with *in situ* hybridizations [48, 49]. This has also been confirmed at the protein level on multiple occasions using Western blotting and, to some extent, at the functional level using electrophysiology and pharmacology (Fig. 2.3). While the GluN1 subunit is ubiquitously expressed in the CNS during embryonic, postnatal development, and throughout adulthood, the GluN2 subunits, as well as GluN3 subunits, differ strikingly in their spatial and temporal expression profile. During early development, GluN2B has the highest level of expression of all GluN2 subunits throughout the CNS. This expression peaks during this second week of postnatal development, but then declines steadily and becomes restricted to the forebrain in adulthood. Similar to GluN2B, the expression of GluN2D peaks during the second week after birth, with a wide distribution at this time (including in the forebrain), before it declines again to weaker expression levels restricted to the brainstem and diencephalon in adulthood. In contrast, GluN2A and GluN2C are barely detectable before birth which drastically changes during the first 2 weeks of postnatal development. The levels of GluN2A rapidly increase during this period and it becomes the predominant GluN2 subunit in the entire CNS in

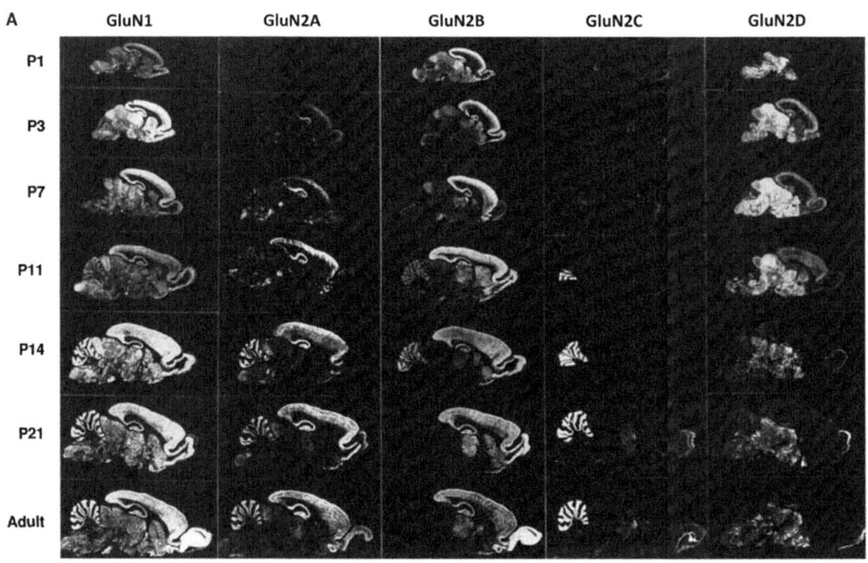

Fig. 2.3 In situ hybridizations showing the distribution of NMDAR GluN2 subunits mRNA in sagittal sections of rat brain over time (birth day (P1) to adulthood). Note the ubiquitous expression of GluN1 at any time in development and adulthood, consistent with the fact that it is an obligatory subunit required for the formation of a functional NMDAR. Adapted from Akazawa et al. [48]

adulthood. The GluN2C subunit appears later in postnatal development (~P10) as well but its expression remains restricted to the cerebellum and the olfactory bulbs throughout adulthood. Among the GluN3 subunits, GluN3A is expressed at low levels during embryonic development, peaks during the first week of postnatal development when its expression is surprisingly widespread, and decreases to low levels in adulthood. Finally, while little attention was dedicated to the GluN3B subunit until recently, its expression profile suggests that it could be central to the physiology of NMDARs in the CNS. Indeed, its expression increases slowly after birth to reach peak and ubiquitous expression in the adult CNS in a similar manner to that of the obligatory GluN1 subunit.

At the functional level, the change in expression profile of NMDAR subunits throughout development is accompanied by changes in kinetics, magnesium-block properties, Ca^{2+} permeability and sensitivity to allosteric modulators of NMDAR-mediated currents (Table 2.1). A typical case of this spatiotemporal maturation of NMDAR subunit expression is the developmental 'switch' from GluN2B to GluN2A in the forebrain; that is, the expression of GluN2B subunits declines as that of GluN2A increases, which has led to the idea that GluN2B-containing NMDARs are replaced by GluN2A-containing NMDARs through postnatal development in the forebrain. This 'GluN2B/GluN2A developmental switch' is a well-accepted concept that prevails in the area CA1 of the hippocampus in particular at the canonical CA3-CA1 synapse. Surprisingly, it has not been fully characterized at the electrophysiological level. But it can still be validated by gathering data from individual studies that have assessed the contribution of GluN2B-homodimers to NMDAR-mediated synaptic currents at the CA3-CA1 synapses at different ages through the use of specific antagonists (Fig. 2.4). Overall it is established that, in adults, GluN2B-NMDARs are absent from CA3-CA1 synapses while GluN2A-homodimers are predominant, such that the population of NMDARs comprises of up to 75 % of GluN2A-NMDARs and about 25 % of GluN2A/GluN2B triheteromers (made of 2 GluN1, one GluN2A and one GluN2B). Therefore, at CA3-CA1 synapses, the replacement of GluN2B by GluN2A is almost total and GluN2B subunits only contribute to the formation of triheteromeric receptors. GluN2A-NMDARs were also found to be the predominant synaptic NMDAR subtype in neurons located in the substantia gelatinosa region of adult rat spinal cord, based on the decay time of synaptic currents, apparent K_d for magnesium and ifenprodil insensitivity [25] .

This spatiotemporal shift in subunit expression is supported by the developmental profile of the intercellular partners involved in anchoring NMDARs at synapse, in particular SAP-102 and PSD95 ([42] and see Sect. 2.2 above). Indeed, SAP-102, which is thought to preferentially interact with the cytoplasmic tail of GluN2B subunits, is enriched at synapses during early stages of development. In the first weeks following birth, however, the expression of PSD-95, which preferentially interacts with GluN2A, increases. Eventually PSD-95 replaces SAP-102 at synapses due to a competition for insertion into the PSD [17, 44, 50, 51].

Unfortunately, biology does not comply with our need for simplicity and the replacement of GluN2B- with GluN2A-NMDARs at synapses is nothing but the exception of a few particular examples of synapses or brain regions and not a gen-

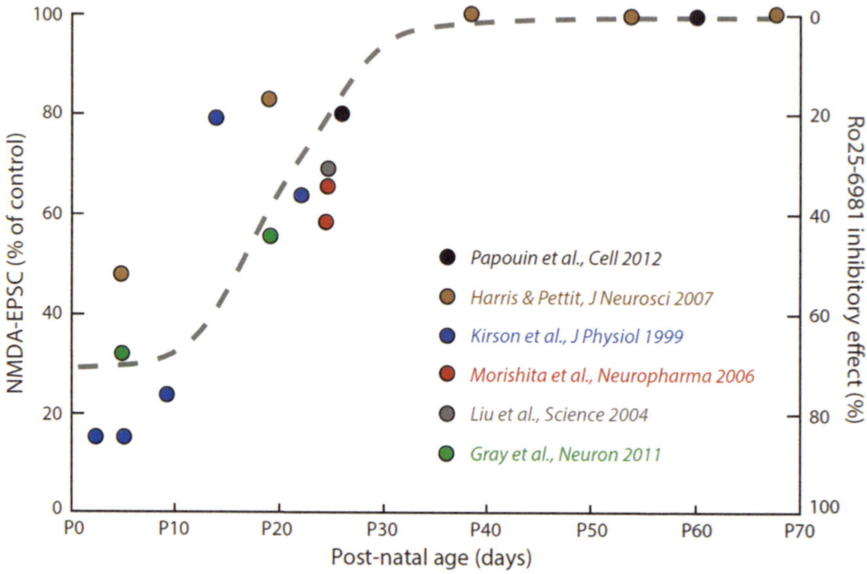

Fig. 2.4 Developmental profile of the sensitivity of NMDAR-mediated synaptic responses to the GluN2B-NMDAR specific antagonist Ro25-6981 at the canonical CA3-CA1 synapse in rat hippocampal acute slices, from data found in the literature. While GluN2B-NMDARs compose most of the population of synaptic NMDARs before P10, they are functionally absent from synapses past P40

eral rule. Indeed, GluN2B-NMDARs represent a major portion of synaptic NMDARs in many other areas of the adult CNS, including in the cortex [52] or in the spinal cord lamina I [53] where the contribution of GluN2B-NMDARs and GluN2D-NMDARs to the synaptic population far exceeds that of GluN2A-NMDARs. Additionally, the NMDAR-subtype content of synapses, in particular GluN2A- and GluN2B-NMDAR content, can be regulated by synaptic activity [3]. Indeed, it was shown that the GluN2B-subunit content is altered following long-term potentiation or depression of glutamatergic synapses in the hippocampus [54], indicating that GluN2B-NMDArs can contribute to synaptic NMDAR signaling in adult even at CA3-CA1 synapses.

2.3.2 *Subunit Composition of Extra-Synaptic NMDARs Is Unresolved*

Given that extra-synaptic NMDARs represent as high as ¾ of the population of NMDARs in the immature hippocampus [14] and an estimated 1/3 of the total NMDAR population on adult hippocampal neurons [12, 17], it is tempting to speculate from 'global' *in situ* hybridization and Western blotting studies that extra-synaptic receptors undergo developmental modifications of their subunit content as well. This, however, has never been addressed, probably because probing extra-synaptic NMDARs in brain slices has proven challenging due to the fact that there is no easy way to directly stimulate and record from this population of receptors.

Nonetheless, it is believed that extra-synaptic NMDARs are enriched in GluN2B-NMDARs in the adult hippocampus, in contrast with the GluN2A-NMDAR content at synapses. This fits with the idea that extra-synaptic NMDARs constitute a distinct population mediating a different function than synaptically located NMDARs. This idea is also quite satisfying when one considers that GluN2B-NMDARs have a higher affinity for glutamate, which is present at lower concentrations in the extra-synaptic space [3, 4, 55]. Additionally, this is consistent with the fact that GluN2B subunits, as well as SAP-102, while not predominant at synapses, are still abundantly expressed by hippocampal neurons in the adult [42–44]. These observations together strongly fuel the idea that GluN2B-NMDARs exist at extra-synaptic sites more abundantly than at synapses. Using the same kind of evidence and reasoning it is believed that GluN2D/C subunits are present at extra-synaptic sites on principal neurons in the cortex and spinal cord [25]. However, the subunit composition of extra-synaptic NMDARs remains a mystery in most brain regions.

Just like different NMDAR subtypes can be segregated at different synapses on a same neuron, it is certainly wise to hypothesize that extra-synaptic NMDARs do not form one homogeneous population at the surface of an entire neuron as far as subunit composition is concerned. It would be interesting to investigate whether, for example, extra-synaptic NMDARs located on apical dendrites share the same composition and properties as those located on distal dendrites, whether different subunit-compositions prevail at the different extra-synaptic specializations (i.e., neuron-glia, or axo-dendritic appositions), whether subunit composition at extra-synaptic sites can be regulated by activity, and whether subunit composition of extra-synaptic NMDARs undergoes a reorganization throughout postnatal development. Additionally, with the development of better subunit-specific antagonists it will soon be possible to reliably address the involvement or GluN2C-, GluN2D- and even GluN3-containing NMDARs in various processes, and it would be interesting to characterize better the extent to which these subtypes participate to extra-synaptic NMDAR signaling.

2.4 Gating of Synaptic and Extra-Synaptic NMDARs

2.4.1 Agonists of the Glutamate-Binding Site of Synaptic NMDARs

Glutamate is the endogenous agonist that binds to the agonist-binding site of synaptic NMDARs, which is located on the GluN2 subunit. It is estimated that glutamate concentration exceeds 3 mM within the first microseconds following a single synaptic release event [15] (Hamilton and Attwell). Considering that the dissociation constant of NMDARs for glutamate is in the micro-molar range, it can be calculated that 97 % of synaptic NMDARs will bind the two molecules of glutamate required for their activation during the first hundreds of microseconds following the release of a single vesicle of glutamate. This means that synaptic NMDARs are fully saturated during a synaptic transmission event. Therefore the rules that govern the activation of the glutamate-binding site of synaptic NMDARs follow a binary probability: 0 in the absence of synaptic activity, 1 during synaptic transmission. This simple fact, though often ignored, highlights the tremendous importance of (1) the existence of another agonist-binding site controlled by a distinct transmitter (D-serine or glycine, see Sects. 2.4.3 and 2.4.4 below) to regulate the gating of NMDARs and (2) the Mg^{2+} block of the pore of the ion channel that, independently the activation of the receptor *per se*, will dictate whether or not an ion flux is permitted through the channel.

Interestingly, some evidence has recently revived the idea that aspartate (D- and L-) could be an endogenous ligand of NMDARs as well. Similar to glutamate, it is synthesized in the CNS, seemingly found in synaptic vesicles released by nerve terminals upon depolarization and taken up after transmission events [56]. Deletion or inhibition of the D-aspartate catabolism enzyme increases the levels of endogenous D-aspartate by 10–20-fold, and mice knocked out for this enzyme have elevated levels of the *N*-methyl derivative of D-aspartate (i.e., NMDA) and enhanced NMDAR-dependent functions such as LTP and spatial learning. While this is by no means a proof that D-, L- or *N*-methyl aspartate are endogenous ligands of NMDARs under normal conditions, this has generated a craze, over the past 15 years, for the idea that aspartate and glutamate could be stored and released from the same vesicle, or that aspartate could have its own vesicular loading and exocytotic pathway, and directly contribute to the activation of NMDARs at glutamatergic synapses. This is supported by the finding that a sialic acid transporter, closely related in sequence to the vesicular glutamate transporters (VGluTs), can transport both glutamate and aspartate [57, 58]. Interestingly, while many vesicular transporters can indeed package several transmitters in the same vesicle, VGluTs do not transport or even recognize aspartate [59, 60]. These observations predicted that, in the absence of VGluTs, excitatory synapses would only signal using synaptic vesicles that contain aspartate; which would permit the activation of NMDARs while AMPAR-mediated transmission would be suppressed. However, it was shown in 2015 by the group of Roger Nicoll that in VGluT1 knock-out mice, NMDAR-mediated synaptic responses were virtually absent and the AMPA/NMDA ratio (whenever measurable) was unaltered. This is consistent with the traditional view that glutamate fully accounts for the activation of synaptic NMDARs and suggested that if aspartate is released at synapses it is present at concentrations too low to have any physiological relevance. Interestingly, however, these data do not exclude the possibility that aspartate could signal to extrasynaptic NMDARs or pre-synaptic NMDARs.

2.4.2 *Agonists of the Glutamate-Binding Site of Extra-Synaptic NMDARs*

While synaptic NMDARs are only activated by brief and acute (i.e. phasic) release of glutamate from the presynaptic terminal, extra-synaptic NMDARs seem to be exposed to different sources of agonist and types of release. The existence of a tonic current mediated by NMDARs was reported in principal neurons of the hippocampus in 1989 [61]. This tonic current is mediated by receptors located outside of synapses and occurs due to sufficient concentrations of ambient agonist, of non-synaptic origin [61–64]. Indeed, this tonic current persists when synaptic activity is suppressed with TTX and, conversely, synaptic NMDAR activity remains intact after the NMDAR-mediated tonic current is blocked with MK-801 [64]. This establishes those receptors as distinct and necessarily distant from synaptic NMDARs. To date, the origin of the ambient glutamate that allows such tonic activation of extra-synaptic NMDARs is still unclear and whether this is indeed glutamate that allows this tonic activation has not, in fact, been demonstrated. Aside from the observation that it is of non-synaptic origin [62, 64] little evidence exist for the source or the nature of the agonist responsible for the NMDAR-mediated tonic current. It is known that the regulation of extra-synaptic glutamate concentration heavily relies on active uptake by astrocytes. Indeed, the glutamate transporter GLT1 is predominantly present on these glial cells and mediates 95 % of the glutamate uptake. Interestingly, the amplitude of NMDAR-mediated tonic current is strongly enhanced either by blocking GLT1 [65, 66], which enables glutamate spill over from synapses, or when slices are challenged with glial toxins [65]. Additionally, a recent study showed that the extent of NMDAR-mediated tonic current depends on glial coverage of synapses in the supra-optic nucleus of the hypothalamus [65]. Therefore, an interesting possibility is that astrocytes could be a key regulator of NMDAR-mediated tonic current by either confining synaptic space with glial processes, thus maintaining the synaptic/extra-synaptic compartmentation, or by modulating ambient glutamate levels.

The existence of an extra-synaptic NMDAR-mediated tonic current implies that a subset of extra-synaptic NMDARs is constitutively activated. However, extra-synaptic NMDARs can also be recruited experimentally by exogenous applications of NMDA or glutamate. This indicates that another subpopulation of extra-synaptic NMDARs exists that is normally silent and can be activated by phasic, non-synaptic release of glutamate. The main manifestation of such receptors are "slow inward currents" (SICs), characterized by a slow decay-time (on the order of seconds), low frequency of occurrence (~0.05 Hz) and a large amplitude (hundreds of pA). They occur spontaneously and have been recorded from principal neurons of the CA1 region of the hippocampus [67–69] and in the superficial layers of the dorsal horn of the spinal cord [70]. Like NMDAR-mediated tonic current, SICs persist in the absence of neuronal and/or synaptic activity [67]. They are also completely suppressed by glial inhibitors [70] and evidence suggests that they are caused by glutamate release from astrocytes onto extra-synaptic NMDARs [22, 55, 68, 71, 72]. It is easy to hypothesize that such currents could result from a "synaptic-like" release of glutamate from glial processes at specialized neuron–glia appositions such as those described above Sect. (2.1.3 and Fig. 2.1).

Challenging view: Beside their mechanistic relevance, these observations are interesting because they open the possibility that extra-synaptic NMDARs may exist as two functionally distinct pools (1), tonically activated receptors, which face supra-threshold amounts of glutamate and co-agonist and (2), overall silent extra-synaptic NMDARs, which can be acutely recruited by exogenous or endogenous agonist release (similar to synaptic NMDARs). If true, this would mean that glutamate availability is not homogenous outside of synapses and could be spatially regulated, giving rise to different sub-compartments within the extra-synaptic space, like suggested by the distribution of extra-synaptic NMDARs themselves. Although this is purely speculative, we propose that it could be relevant to consider the extra-synaptic space as being comprised of separate compartments instead of one large homogenous volume, at least from the point of view of glutamate and NMDARs.

2.4.3 Agonists of the Glycine-Binding Site of Synaptic NMDARs

In 1987 a simple, yet revolutionary observation was published in a Nature article and described in those words: "In cultured neurons, the magnitude of the whole cell current produced by NMDA appeared to depend on the speed of perfusion of NMDA-containing solutions: slower movement of the perfusion solution resulted in larger response." The authors hypothesized that "a substance that augments the response to NMDA is tonically released from the cultured cells (neurons or glia), and accumulates when the perfusion is slow" [73]. Johnson and Ascher had just made the discovery that NMDARs require two agonists for their activation. Their investigation led them to individually screen the effect of amino acids present in the cultured medium, and they made the discovery that "the effects of conditioned medium were reproduced by glycine." Indeed, the main finding of this work Fig. (2.4) is that application of NMDA alone (or glutamate) on outside-out excised patches does not permit the opening of the NMDAR ion channel, and neither do applications of glycine alone. When both are co-applied however, NMDARs are activated and this results in sustained currents flowing through the receptor's ion channel Fig. (2.5). While authors concluded that "glycine augmented the response to glutamate as well as that to NMDA" the correct conclusion of this experiment is that glycine *allowed* the response to glutamate as well as that to NMDA [74]; and this finding is the first demonstration that NMDARs harbor two distinct agonist-binding sites that recognize, and are activated by, two different ligands. As a concluding remark, authors hypothesized that "the glycine-binding site and the NMDA-binding site are on two distinct proteins [i.e. subunits]", which was confirmed the following year in a Science publication by Kleckner and Dingledine who demonstrated that the glycine-binding site is on the GluN1 subunit while the

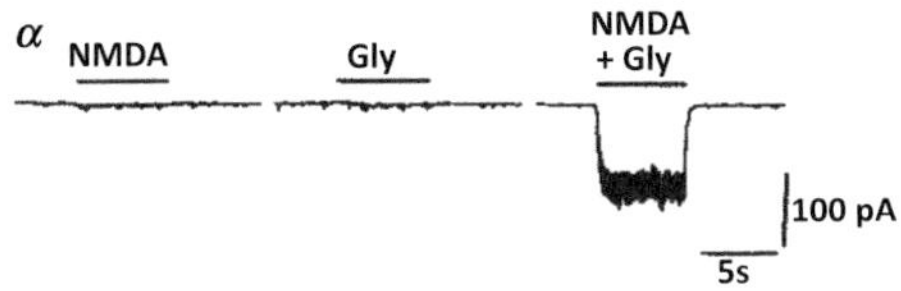

Fig. 2.5 Effect of glycine (1 μM) on the current induced by NMDA (10 μM) on outside-out excised patches containing NMDARs. Note that applications of NMDA or glycine alone do not elicit currents, consistent with the fact that the simultaneous binding of the two agonists on their respective agonist-binding site is required for the activation of NMDAR. From [73]

glutamate-binding site is on the GluN2 subunit. Since then, the agonist binding domain of the GluN1 subunit of NMDARs has been coined the "glycine-binding site" and glycine was termed "co-agonist" of NMDARs. The idea that glycine, an abundant and ubiquitous amino acid present in culture medium and in the extracellular space of the brain parenchyma, is the endogenous agonist of the glycine-binding site remained unchallenged for nearly 20 years. It was supported by the existence of glycine transporters (GlyT1) on glial processes that surround synaptic contacts and mediate a powerful uptake of glycine. This was thought to regulate the concentration of glycine at excitatory synapses, providing a powerful means to modulate the degree of activation of the NMDAR glycine-binding site.

This canonical view was challenged by the groups of Salmon Snyder and Stephane Oliet, first in 1995 and then in 2006, respectively. After the discovery that D-amino acids exist in mammalian CNS [75–77], the observation was made by Schell et al. [78] that the localization of D-serine throughout the brain resembles that of NMDARs, and the binding pattern of (^{3}H)D-serine revealed by autoradiography strongly resembled that of (^{3}H)glycine binding. This strongly suggested that D-serine is an endogenous ligand for the glycine-binding site of NMDARs, an idea that received very little attention from most of the neuroscience community—with the notable exception of the *glial* community. Indeed, as the study of glial cells had started to thrive, the publication by Schell et al. showed outstanding potential contained in this statement: "We were surprised to observe under high magnification that D-serine immunoreactivity was associated exclusively with glia".

The first demonstration that D-serine, and not glycine, is the endogenous ligand of the glycine-binding site of the NMDAR in brain tissue came from Panatier et al. [79], and echoed similar earlier demonstrations carried out in cultures by the Snyder group. Soon after, it was established that D-serine is the endogenous co-agonist of NMDARs at many, but not all, central synapses (reviewed in Papouin et al. [80]), such as in layer 5 of the cortex, in the retina, in the amygdala, in the nucleus accumbens, in the spinal cord and, notably, at the canonical CA3-CA1 synapses of the hippocampus [13]. The strength and impact of this finding are greatly underappreciated. Indeed, given that the gating of NMDARs by glutamate at synapses is binary (all or none, see Sect. 2.4.1), and given the importance of NMDARs in the development, physiology and pathology of the CNS, the notion that glial cells are responsible for the supply and regulation of the glycine-binding site's agonist makes glial cells the primary regulator of the most investigated receptor in neuroscience. This also means that astrocytes are directly involved in processes such as synaptic plasticity and learning. Additionally, with its glutamate-binding site controlled by the pre-synaptic neuron, its glycine-binding site controlled by astrocytes, and its magnesium block reliant on the activity of the post-synaptic neuron, the NMDAR can be viewed as a molecular-scale model of the tripartite synapse and thus a remarkable tool to study it. The rules that govern the release of D-serine are still incompletely elucidated and subject to debates that spread beyond the scope of this chapter.

That D-serine is the endogenous ligand of the glycine-binding site of synaptic NMDARs is not true at every synapse in the CNS. Many instances have been found where the historical co-agonist glycine happens to be the endogenous ligand of the glycine-binding site. Such is the case in the spinal cord [81], on retinal ganglion cells [82] and in the nucleus of the tractus solitarius [79]. In structures where glycinergic innervation is abundant, such as in the retina and the spinal cord, it was shown that glycine that serves as an endogenous co-agonist of NMDARs at glutamatergic synapses [81, 82] originates from nearby inhibitory synapses. It is thought that following inhibitory synapses activity, glycine spills over and diffuses to bind to NMDARs at remote excitatory synapses. This highlights an interesting cross-talk between inhibitory and excitatory synapses in these regions along with the potential for inhibitory synapses to impact synaptic plasticity occurring at neighboring excitatory terminals.

Challenging view: When it was discovered in 1987 that NMDARs bind another ligand, and that this second-agonist is required to permit the activation of the receptor by glutamate, it was very naturally (but unfortunately) termed "co-agonist" of NMDARs. This misnomer has resulted in a multitude of inaccurate views and descriptions of NMDAR co-agonist function, ranging from "helper" for receptor activation to "allosteric modulator". Not only is the terminology used in these examples erroneous, but the view they attempt to describe about the function of the co-agonist is wrong as well. In this case, like often in science, inspection of the original publication by Johnson and Ascher proves to be enlightening as it reveals that the so-called NMDAR co-agonist is nothing less than a full agonist at the receptor: (1) it binds to its own, separate, ligand-binding site on the GluN1 subunit, (2) it is absolutely required for the opening of the ion pore and (in the presence of glutamate) is sufficient for this activation, and (3) the efficacy of glycine and D-serine at the GluN1 agonist-binding site is ~100 %, which means they are full, not partial, agonists. In other words, NMDAR possesses two distinct agonist-binding sites, one harbored by each GluN1 subunit that binds glycine/D-serine, and one harbored by each GluN2 subunits that binds glutamate; in such a way that glutamate and glycine/D-serine play the exact same function on different subunits. The only reason why glycine was termed "co-agonist", while glutamate is termed "agonist", is because its role in controlling the activation of NMDARs was discovered after that of glutamate, but from a biological and molecular point of view, there is no such thing as a *main* or *primary* agonist of NMDARs. These considerations are also well illustrated by the simple fact that GluN3-NMDARs (made of two GluN1 and two GluN3 subunits that bind glycine/D-serine but not glutamate) lack a glutamate-binding site and are therefore only activated by glycine or D-serine while insensitive to glutamate.

2.4.4 Co-agonists of the Glycine-Binding Site of Extra-Synaptic NMDARs

The identity of the co-agonist of extra-synaptic NMDARs has only been addressed once, by our group [80]. We showed that, on apical dendrites of CA1 pyramidal neurons, extra-synaptic NMDARs are gated by glycine, and not D-serine. In striking contrast, NMDARs located at CA3-CA1 synapses on the same neurons are gated by D-serine, not glycine, like mentioned earlier. It is reasonable to assume that the availability of glycine and D-serine are regulated by distinct mechanisms. Therefore, on CA1 pyramidal neurons, the gating of the co-agonist binding site of NMDARs contributes to segregating synaptic and extra-synaptic NMDARs as two functionally distinct pools. Whether glycine is also the co-agonist of extra-synaptic NMDARs located elsewhere on CA1 neurons, and whether this applies to other brain areas, remains to be addressed. If the identity of the co-agonist is used as a way to more efficiently separate synaptic from extra-synaptic NMDARs as a general rule, then one might expect that wherever D-serine is the co-agonist of synaptic NMDARs, glycine would gate extra-synaptic NMDARs, and vice versa.

The origin of glycine available to extra-synaptic NMDARs is unclear, especially in the hippocampus. In this structure, the presence of glycinergic terminals has never been established (the inhibitory transmission is entirely abolished by GABA receptor antagonists) and the expression of functional glycine receptors (GlyRs) is thought to stop after birth [83–85], even though the existence of extra-synaptic GlyRs has been documented in adults [86, 87]. Yet, *in vivo* microdialysis reported amounts of free glycine as high as 10 μM in the hippocampus [88, 89]. *In vivo*, a major source of extracellular glycine in the CNS could be the blood flow since approximately 200 μM of glycine are found in the blood [90] and since glycine is able to cross the endothelial wall of capillaries by means of glycine transporters [91, 92]. In slices, however, blood vessels are emptied of their initial content, indicating that a source of glycine exists in the brain parenchyma itself. With an average intracellular glycine concentration of 3–6 mM [93], glial cells could be a major source of glycine in brain slices. In the hippocampus, it was also proposed that glycine could be co-released with GABA at inhibitory synapses by interneurons [94] similar to what occurs in the thalamus, the brainstem, the spinal cord and the cortex [95, 96]. Notably, that glycine is present at 10 μM in brain parenchyma would suggest that the amounts of glycine available to extra-synaptic NMDARs are high enough to be saturating [88, 89], especially if those receptors are GluN2B-NMDARs ($K_d < 1$ μM). This would leave little room for the modulation of NMDAR activity through the co-agonist-binding site at extra-synaptic locations. Although surprising, this idea is supported by observations that exogenous applications of co-agonist do not enhance the amplitude of NMDAR-mediated tonic current [64].

2.4.5 Endogenous and Exogenous NMDARs Allosteric Modulators

Several competitive antagonists, binding directly onto the agonist-binding site of either the GluN1 subunit or the GluN2 subunit where they compete with the endogenous agonist and prevent or reduce its conformational activation of the site, are available, efficient and specific for NMDARs if used at appropriate concentrations. This includes the glycine-binding site antagonist 5,7-Dichlorokynurenic acid (DCKA) and the glutamate binding-site antagonist D(−)-2-Amino-5-phosphonopentanoic acid (D-AP5). Blockers of the ion channel such as Dizocilpine (MK-801) also exist that act as non-competitive antagonists since they alter the permeability of the channel without changing the activation rules of the receptor. Beside these competitive and non-competitive antagonists, a number of endogenous and exogenous compounds can modulate the activity of NMDARs by binding on specific sites that are distinct from the agonist binding sites or the ion channel. They act by changing the conformation of the receptor which usually modifies the open channel probability (*P*o) but, contrary to agonists and competitive antagonists, they do not interfere with the activation of the receptor. Among the endogenous allosteric modulators, the most unacknowledged of them all is also the simplest and the most important: H^+ ions (protons). Protons bind to the NMDAR at an unknown location thought to be closely associated with the channel gate. Proton binding impacts the N-terminal domain (NTD) of the GluN2 subunit and trigger a "closed" conformation of this domain which favors the closure of the ion pore of NMDAR channel. Therefore, H^+ binding reduces the *P*o of the channel [97]. This is of the uttermost importance mainly for two reasons: (1) The EC_{50} of the proton-binding site for protons is in the range of physiological pH (~6.9 for GluN2A and 7.5 for GluN2B subunit) which means that protons exert a tonic and basal inhibition of NMDAR channel *P*o and that this inhibition can differentially affect particular NMDAR-subtypes for subtle pH changes. (2) The inhibitory effect of H^+ is directly responsible for the subunit-selectivity and efficacy of other allosteric modulators such as zinc (Zn^{2+}) and Ifenpodil/Ro25-6981. Indeed these competitive antagonists act by causing conformational changes in the receptor that increase the sensitivity of the proton-binding site and thus enhance the inhibition by H^+, therefore further reducing the *P*o of the channel. Zinc, like protons, is an endogenous allosteric modulator of NMDARs that binds in the NTD of GluN2A subunits with a remarkable selectivity since its affinity for GluN2A-NTD is 1000 times better than that for any other subunit. Under physiological pH, zinc only produces a ~70 % inhibition of GluN2A-NMDAR mediated current. Interestingly, zinc is co-stored in vesicles and co-released with glutamate. While, ambient extracellular zinc levels are too low to inhibit synaptic GluN2A-NMDARs in a tonic manner, sustained synaptic transmission elicits a surge of zinc concentration at excitatory synapses that causes an endogenous subunit-specific inhibition of NMDAR transmission that is thought to mediate an activity-dependent regulation of neuronal circuits [98].

Surprisingly, 40 years after the discovery of the first NMDAR antagonist, the number of pharmacological reagents available to discriminate between NMDAR-subtypes is still very limited as reviewed in Neyton and Paoletti [99]. In particular, antagonists selective for GluN2C-, GluN2D- and GluN3-containing receptors as well as specific antagonists for NMDAR tiheteromers, such as GluN2A/GluN2B-NMDARs, are still lacking. Yet, a large number of compounds are available online from various chemical suppliers that are described as subunit-selective based on their *tendency to prefer* a particular NMDAR-subtype. Unfortunately, while some of them indeed have a higher affinity for a particular NMDAR subunit or subtype, their *affinity* is not high enough to allow a full inhibition of this subtype/subunit without impacting other subtypes/subunits. This is typically the case of PPDA, often presented as a GluN2C- and GluN2D-specific antagonist based on its preferential binding to these subunits (Ki ~ 0.1 μM). Unfortunately, it Ki in the very same order of magnitude for the other subunits (GluN2B: 0.3 μM; GluN2A: 0.6 μM), such that PPDA is not a subunit-selective antagonist specific for GluN2C/GluN2D-containing NMDARs. The same has been established for NVP-AAM077 [100], a putative GluN2A-NMDAR specific antagonist. As a general rule, it is safe to consider that an antagonist is only selective for a NMDAR subtype/subunit if its affinity for the latter is at least an order of magnitude higher than that for other subtypes/subunits [99].

2.5 Functions of NMDARs in Relation to Their Location

While the general role of NMDARs in various CNS functions is clearly establishes, the precise contribution of synaptic vs. extra-synaptic receptors to each of them remains mostly elusive. Some studies have reported a role for non-synaptic NMDARs in extra-synaptic inhibition [101] and dendritic dynamic range compression [102], but the main functions in which extra-synaptic NMDARs have been involved and studied are synaptic plasticity, excitotoxicity/ischemia, excitability, and neuron-glia interactions.

2.5.1 Synaptic Plasticity, a Matter of Subunits or Location?

Using genetic and pharmacological approaches, there has been intense speculation about the role of specific NMDAR subtypes in the selective induction of LTP and LTD [103–108]. It was proposed that GluN2A-NMDARs preferentially trigger LTP, whereas GluN2B-NMDARs are preferentially associated with LTD. Given the very complex and diversified expression pattern of NMDAR subunits throughout the CNS and over time, this tempting proposal seemed improbable. While it might be true in specific cases, this dichotomy has indeed been significantly and repeatedly challenged [104, 106] and the idea that a particular subtype of NMDAR is specifically involved in inducing potentiation or depression at glutamatergic synapses seems over-simplifying, as reviewed in great detail in [3]. Since synaptic and extra-synaptic NMDARs often seem to have distinct subunit compositions, this controversy raised the question of whether NMDAR location, rather than subunit composition *per se*, could constitute a determining factor to the direction of synaptic plasticity and be the biological reason for discrepancies between studies [109]. While this interesting idea was formulated 10 years ago, the role of extra-synaptic NMDARs in LTP and LTD has rarely been characterized directly, i.e. independently of subunit composition. Because there is no simple rule linking the cellular location of NMDARs to their subunit composition, and because the subunit content of receptors located outside of synapses is unresolved, it is difficult to interpret results involving extra-synaptic NMDARs in synaptic plasticity based on subtype-specific pharmacological experiments which, unfortunately, is the vast majority of studies on that matter. Activating or silencing specifically synaptic or extra-synaptic NMDARs thus requires different approaches that do not rely on subunit composition. The main alternative is the use of the open-channel blocker MK-801, which allows inactivation of NMDARs that are recruited while leaving the silent/not recruited receptors intact. When combined with low-frequency stimulation of afferent fibers, this method allows the selective blockade of synaptic NMDARs and leaves most of their extra-synaptic counterparts intact [12, 13, 64]. This method presents the advantage that it only relies on whether receptors are active or not during MK-801 application, and therefore circumvents caveats associated with the use of subunit-selective reagents. Interestingly, using such an approach, Xu et al. [110] found that selective stimulation of extra-synaptic NMDARs triggers LTD in CA1 neurons. Another approach is based on the discovery that D-serine and glycine gate synaptic and extra-synaptic NMDARs respectively [13]. Taking advantage of this segregation, we showed in adult rat hippocampal slices that synaptic NMDARs, but not extra-synaptic ones, are required for LTP induction, whereas activation of receptors at both locations is required for LTD. These findings thus confirm the idea proposed by Dr. Rusakov [109] that what matters in LTP and LTD induction may be the location, rather than the subtype, of NMDARs recruited. If the compartmentalization of NMDAR by their co-agonist gating was to be generalized to other brain regions, this might be a very powerful and convenient way, in addition to the MK-801 approach, to study the role of synaptic and extra-synaptic NMDARs in various functions independently of their composition.

2.5.2 Is Excitotoxicity Caused by Extra-Synaptic NMDARs?

Because NMDARs are highly permeable to calcium, they not only relay a physiological signal to neurons, but can also trigger intracellular signaling cascades leading to cell death. In fact, the link between excitotoxicity [111] and excessive glutamate release [112] were described years before synaptic plasticity. NMDARs were demonstrated to be the main source of calcium responsible for glutamate-induced excitotoxicity in the late 1980s [113–115]. Ever since, NMDARs have been renowned for their dual roles in physiology and pathophysiology, and the mechanisms of NMDAR-induced cell death is one of the most investigated aspects of this receptor.

In 2002, Hardingham et al. published the first compelling evidence that synaptic and extra-synaptic NMDAR activation have opposing effects on cell fate. Ever since, the prevailing theory is that the activity of synaptic NMDARs favors neuronal survival, via the phosphorylation of intracellular factors such as CREB or Erk1/2. In contrast, cell death is mainly mediated by the activation of extra-synaptic NMDARs (mostly GluN2B-NMDARs in this study) which inhibits the CREB and Erh1/2 pathways in addition to directly promoting pro-death signaling via caspase-3 [16]. This view gained additional attention as the use of memantine developed. Memantine is thought to be an extra-synaptic NMDARs blocker and was shown to reduce neuronal death in models for neurodegenerative disorders such as Huntington's [116] or Alzheimer's disease [117–119]. However, numerous observations now point to the limitation of the theory that synaptic NMDARs favor survival while extra-synaptic NMDARs promote cell-death. Indeed, this finding, obtained in neuronal cultures, has hardly been replicated in slices or *in vivo*. In fact, several experiments point to a role for synaptic NMDARs in cell death during NMDA application in acute slices or oxygen glucose deprivation (OGD) *in vivo* [13, 80, 120, 121]. Amadoro et al. [122] demonstrated that synaptic NMDARs mediate hypoxia-induced neurotoxicity almost entirely, whereas blocking extra-synaptic NMDARs provides no protection. Similarly, we showed that NMDA-induced excitotoxicity in slices is entirely mediated by synaptic NMDARs in the area CA1 of the hippocampus since it can be prevented by silencing receptors at this location [13]. On the contrary, silencing extra-synaptic NMDARs had no effect on NMDA-induced cell-death. Other recent work [120, 123], following the protocols established by Hardingham et al. [124], failed to replicate the findings that extra-synaptic NMDARs, but not synaptic ones, promote cell-death. This controversy, questioning the role of extra-synaptic NMDARs in excitotoxicity, is also fueled by recent and thorough insights into the mechanism of the use-dependent NMDAR blocker memantine [125]. Memantine was proposed to preferentially block extra-synaptic receptors over synaptic ones, owing to its fast off-rate, low affinity, voltage-dependent binding and uncompetitive nature [16, 125, 126]. The fact that memantine can successfully prevent neuronal death *in vitro* or in pathological conditions [127, 128] has thus strongly contributed to the extra-synaptic hypothesis of excitotoxicity. However, the group of Steven Mennerick demonstrated, in two studies, that memantine has a strong and fast inhibitory effect on synaptic NMDARs (80 % in 5 min) under basal low-frequency stimulation. Additionally, they show that the neuroprotective effect of memantine is mediated by the blockade of synaptic, not extra-synaptic, NMDARs [125] (Katsuki et al.). This body of evidence strongly questions the prevalent role of extra-synaptic NMDARs in triggering neuronal death during excitotoxic conditions in culture, and suggests that this theory does not apply to more intact preparations.

2.5.3 *Neuron-Glia Interactions and Excitability*

The mechanisms and roles of SICs and tonic currents, and how astrocytes play a role in these forms of signaling have been recently reviewed by Balázs Pál (Front Cell Neuro [129]).

It is accepted that the tonic current mediated by extra-synaptic NMDARs plays a role in the excitability of principal neurons and in modulating dendritic inputs [130]. This might be particularly true for interneurons [131] which excitability is crucial in setting synchronized neuronal population dynamics. It also becomes increasingly clear that NMDAR-mediated tonic current can be upregulated in pathological conditions, ranging from cocaine addiction [132] to Alzheimer's disease [117]. However, the relevance of such tonic current to neuronal physiology and the impact of its pathological disturbance need further investigation. Notably, even though its exact function is unknown, NMDAR-mediated tonic current certainly plays a major role in neuronal and network physiology because it has been observed in virtually every brain region thus far.

SICs were shown to result from the release of glutamate by astrocytes onto neuronal extra-synaptic NMDARs, and to occur simultaneously in distinct neurons within 100 μm of each other [67], which fits with the idea that an astrocyte's anatomical territory spans an area approximately 50–100 μm in diameter. Based on these observations, SICs have been proposed to play a role in neuronal synchrony and network excitability [67, 81] but this aspect still remains poorly understood and would require more investigation. In addition, since they have only been observed in slices so far, often under non-physiological conditions (low magnesium, GLT1 blockers), the question of whether SICs represent an important feature of neuron–glia signaling that also occurs *in vivo* remains to be established. Based on the observation that extra-synaptic NMDARs are often located at seemingly specialized neuron-astrocyte contacts, it might be relevant to investigate SICs as the manifestation of an extra-synaptic, yet synaptic-like, astrocyte-to-neuron form of signaling.

2.6 Concluding Remarks

In conclusion we would like to draw readers' attention to the fact that the subcellular location of the NMDAR has emerged as a key determinant of NMDAR-mediated physiological functions as well as NMDAR-related pathological conditions. Such location appears to be equally or potentially more important than NMDAR subunit-composition. Furthermore, extreme methodological and intellectual caution should be used when employing subunit-based pharmacological approaches to decipher the role of a particular subset of NMDAR in a given context. Not only are some of the available reagents not selective enough, but also the subunit composition of NMDARs is not conclusively telling of their location since there seems to be no clear or absolute subunit-hallmarks that differentiate synaptic from extra-synaptic NMDARs. Finally, many aspects of extra-synaptic NMDARs suggest that they could be involved in a complex and specific type of extra-synaptic cell-to-cell communication. That this facet of NMDAR physiology has remained largely unexplored promises new exiting lines research and future paradigm-shifting discoveries. As a starting point, we propose that the availability of glutamate and co-agonists, and the distribution of NMDARs, are not homogenous outside of synapses and could be spatially regulated, thus delineating distinct functional and morphological sub-compartments within the extra-synaptic space.

References

1. Biscoe TJ et al. D-alpha-aminoadipate as a selective antagonist of amino acid-induced and synaptic excitation of mammalian spinal neurones. Nature. 1977;270:743.
2. Davies J, Watkins JC. Actions of D and L forms of 2-amino-5-phosphonovalerate and 2-amino-4-phosphonobutyrate in the cat spinal cord. Brain Res. 1982;235:378–86.
3. Paoletti P, Bellone C, Zhou Q. NMDA receptor subunit diversity: impact on receptor properties, synaptic plasticity and disease. Nat Rev Neurosci. 2013;14:383–400.
4. Paoletti P. Molecular basis of NMDA receptor functional diversity. Eur J Neurosci. 2011;33:1351–65.
5. Monaghan DT, Jane DE. In: Van Dongen AM, editorBiology of the NMDA receptorBoca Raton, FL: CRC Press/Taylor & Francis; 2009.
6. LeMaistre JL, Sanders SA, Stobart MJ, Lu L, Knox JD, Anderson HD, Anderson CM. Coactivation of NMDA receptors by glutamate and D-serine induces dilation of isolated middle cerebral arteries. J Cereb Blood Flow Metab. 2012;32:537–47.
7. Stobart JL, Lu L, Anderson HD, Mori H, Anderson CM. Astrocyte-induced cortical vasodilation is mediated by D-serine and endothelial nitric oxide synthase. Proc Natl Acad Sci U S A. 2013;110:3149–54.
8. Lee MC, Ting KK, Adams S, Brew BJ, Chung R, Guillemin GJ. Characterisation of the expression of NMDA receptors in human astrocytes. PLoS One. 2010;5:14123.
9. Cao N, Yao ZX. Oligodendrocyte N-methyl-D-aspartate receptor signaling: insights into its functions. Mol Neurobiol. 2013;47:845–56.
10. Chen S, Diamond JS. Synaptically released glutamate activates extrasynaptic NMDA receptors on cells in the ganglion cell layer of rat retina. J Neurosci. 2002;22:2165–73.
11. Clark BA, Cull-Candy SG. Activity-dependent recruitment of extrasynaptic NMDA receptor activation at an AMPA receptor-only synapse. J Neurosci. 2002;22:4428–36.
12. Harris AZ, Pettit DL. Extrasynaptic and synaptic NMDA receptors form stable and uniform pools in rat hippocampal slices. J Physiol. 2007;584:509–19.
13. Papouin T, Ladépêche L, Ruel J, Sacchi S, Labasque M, Hanini M, Groc L, Pollegioni L, Mothet JP, Oliet SH. Synaptic and extrasynaptic NMDA receptors are gated by different endogenous coagonists. Cell. 2012;150:633–46.
14. Tovar KR, Westbrook GL. The incorporation of NMDA receptors with a distinct subunit composition at nascent hippocampal synapses in vitro. J Neurosci. 1999;19:4180–8.
15. Clements JD. Transmitter timecourse in the synaptic cleft: its role in central synaptic function. Trends Neurosci. 1996;19:163–71.
16. Hardingham GE, Bading H. Synaptic versus extrasynaptic NMDA receptor signalling: implications for neurodegenerative disorders. Nat Rev Neurosci. 2010;11:682–96.
17. Petralia RS, Wang YX, Hua F, Yi Z, Zhou A, Ge L, Stephenson FA, Wenthold RJ. Organization of NMDA receptors at extrasynaptic locations. Neuroscience. 2010;167:68–87.

18. Zhang J, Diamond JS. Subunit- and pathway-specific localization of NMDA receptors and scaffolding proteins at ganglion cell synapses in rat retina. J Neurosci. 2009;29:4274–86.
19. Jourdain P, Bergersen LH, Bhaukaurally K, Bezzi P, Santello M, Domercq M, Matute C, Tonello F, Gundersen V, Volterra A. Glutamate exocytosis from astrocytes controls synaptic strength. Nat Neurosci. 2007;10:331–9.
20. Kunz PA, Roberts AC, Philpot BD. Presynaptic NMDA receptor mechanisms for enhancing spontaneous neurotransmitter release. J Neurosci. 2013;33:7762–9.
21. Aoki C, Rhee J, Lubin M, Dawson TM. NMDA-R1 subunit of the cerebral cortex co-localizes with neuronal nitric oxide synthase at pre- and postsynaptic sites and in spines. Brain Res. 1997;750:25–40.
22. Bezzi P, Gundersen V, Galbete JL, Seifert G, Steinhäuser C, Pilati E, Volterra A. Astrocytes contain a vesicular compartment that is competent for regulated exocytosis of glutamate. Nat Neurosci. 2004;7:613–20.
23. Clark BA, Farrant M, Cull-Candy SG. A direct comparison of the single-channel properties of synaptic and extrasynaptic NMDA receptors. J Neurosci. 1997;17:107–16.
24. Kharazia VN, Weinberg RJ. Immunogold localization of AMPA and NMDA receptors in somatic sensory cortex of albino rat. J Comp Neurol. 1999;412:292–302.
25. Momiyama A. Distinct synaptic and extrasynaptic NMDA receptors identified in dorsal horn neurones of the adult rat spinal cord. J Physiol. 2000;523:621–8.
26. Petralia RS, Wang YX, Wenthold RJ. Internalization at glutamatergic synapses during development. Eur J Neurosci. 2003;18:3207–17.
27. Rao A, Craig AM. Activity regulates the synaptic localization of the NMDA receptor in hippocampal neurons. Neuron. 1997;19:801–12.
28. Valtschanoff JG, Burette A, Wenthold RJ, Weinberg RJ. Expression of NR2 receptor subunit in rat somatic sensory cortex: synaptic distribution and colocalization with NR1 and PSD-95. J Comp Neurol. 1999;410:599–611.
29. Petralia RS, Wang YX, Wenthold RJ. NMDA receptors and PSD-95 are found in attachment plaques in cerebellar granular layer glomeruli. Eur J Neurosci. 2002;15:583–7.
30. Fujisawa S, Aoki C. In vivo blockade of N-methyl-D-aspartate receptors induces rapid trafficking of NR2B subunits away from synapses and out of spines and terminals in adult cortex. Neuroscience. 2003;121:51–63.
31. Rosenmund C, Feltz A, Westbrook GL. Synaptic NMDA receptor channels have a low open probability. J Neurosci. 1995;15:2788–95.
32. Tao-Cheng JH, Gallant PE, Brightman MW, Dosemeci A, Reese TS. Structural changes at synapses after delayed perfusion fixation in different regions of the mouse brain. J Comp Neurol. 2007;501:731–40.
33. Takumi Y, Ramírez-León V, Laake P, Rinvik E, Ottersen OP. Different modes of expression of AMPA and NMDA receptors in hippocampal synapses. Nat Neurosci. 1999a;2:618–24.
34. Groc L, Bard L, Choquet D. Surface trafficking of N-methyl-D-aspartate receptors: physiological and pathological perspectives. Neuroscience. 2009;158:4–18.
35. Groc L, Heine M, Cousins SL, Stephenson FA, Lounis B, Cognet L, Choquet D. NMDA receptor surface mobility depends on NR2A-2B subunits. Proc Natl Acad Sci U S A. 2006;103:18769–74.
36. Tovar KR, Westbrook GL. Mobile NMDA receptors at hippocampal synapses. Neuron. 2002;34:255–64.
37. Axelrod D, Koppel DE, Schlessinger J, Elson E, Webb WW. Mobility measurement by analysis of fluorescence photobleaching recovery kinetics. Biophys J. 1976;16:1055–69.
38. Poo M. Rapid lateral diffusion of functional A Ch receptors in embryonic muscle cell membrane. Nature. 1982;295:332–4.
39. Triller A, Choquet D. New concepts in synaptic biology derived from single-molecule imaging. Neuron. 2008;59:359–74.
40. Bard L, Groc L. Glutamate receptor dynamics and protein interaction: lessons from the NMDA receptor. Mol Cell Neurosci. 2011;48:298–307.

41. Bard L, Sainlos M, Bouchet D, Cousins S, Mikasova L, Breillat C, Stephenson FA, Imperiali B, Choquet D, Groc L. Dynamic and specific interaction between synaptic NR2-NMDA receptor and PDZ proteins. Proc Natl Acad Sci U S A. 2010;107:19561–6.
42. Yoshii A, Sheng MH, Constantine-Paton M. Eye opening induces a rapid dendritic localization of PSD-95 in central visual neurons. Proc Natl Acad Sci U S A. 2003;100:1334–9.
43. Sans N, Petralia RS, Wang YX, Blahos II J, Hell JW, Wenthold RJ. A developmental change in NMDA receptor-associated proteins at hippocampal synapses. J Neurosci. 2000;20: 1260–71.
44. van Zundert B, Yoshii A, Constantine-Paton M. Receptor compartmentalization and trafficking at glutamate synapses: a developmental proposal. Trends Neurosci. 2004;27:428–37.
45. Valtschanoff JG, Weinberg RJ. J Neurosci. 2001;21:1211–7.
46. Pérez-Otaño I, Ehlers MD. Homeostatic plasticity and NMDA receptor trafficking. Trends Neurosci. 2005;28:229–38.
47. Steigerwald F, Schulz TW, Schenker LT, Kennedy MB, Seeburg PH, Köhr G. C-Terminal truncation of NR2A subunits impairs synaptic but not extrasynaptic localization of NMDA receptors. J Neurosci. 2000;20:4573–81.
48. Akazawa C, Shigemoto R, Bessho Y, Nakanishi S, Mizuno N. Differential expression of five N-methyl-D-aspartate receptor subunit mRNAs in the cerebellum of developing and adult rats. J Comp Neurol. 1994;347:150–60.
49. Laurie DJ, Seeburg PH. Regional and developmental heterogeneity in splicing of the rat brain NMDAR1 mRNA. J Neurosci. 1994;14:3180–94.
50. Barria A, Malinow R. Subunit-specific NMDA receptor trafficking to synapses. Neuron. 2002;35:345–53.
51. Gladding CM, Raymond LA. Mechanisms underlying NMDA receptor synaptic/extrasynaptic distribution and function. Mol Cell Neurosci. 2011;48:308–20.
52. Thomazeau A, Lassalle O, Iafrati J, Souchet B, Guedj F, Janel N, Chavis P, Delabar J, Manzoni OJ. Prefrontal deficits in a murine model overexpressing the down syndrome candidate gene dyrk1a. J Neurosci. 2014;34:1138–47.
53. Hildebrand ME, Pitcher GM, Harding EK, Li H, Beggs S, Salter MW. GluN2B and GluN2D NMDARs dominate synaptic responses in the adult spinal cord. Sci Rep. 2014;4:4094.
54. Lee MC, Yasuda R, Ehlers MD. Metaplasticity at single glutamatergic synapses. Neuron. 2010;66:859–70.
55. Hamilton NB, Attwell D. Do astrocytes really exocytose neurotransmitters? Nat Rev Neurosci. 2010;11:227–38.
56. Morland C, Nordengen K, Larsson M, Prolo LM, Farzampour Z, Reimer RJ, Gundersen V. Vesicular uptake and exocytosis of L-aspartate is independent of sialin. FASEB J. 2013;27:1264–74.
57. Miyaji T, Echigo N, Hiasa M, Senoh S, Omote H, Moriyama Y. Identification of a vesicular aspartate transporter. Proc Natl Acad Sci U S A. 2008;105:11720–4.
58. Miyaji T, Omote H, Moriyama Y. Functional characterization of vesicular excitatory amino acid transport by human sialin. J Neurochem. 2011;119:1–5.
59. Herring BE, Silm K, Edwards RH, Nicoll RA. Is Aspartate an Excitatory Neurotransmitter? J Neurosci. 2015;35:10168–71.
60. Reimer RJ, Edwards RH. Organic anion transport is the primary function of the SLC17/type I phosphate transporter family. Pflugers Arch. 2004;447:629–35.
61. Sah P, Hestrin S, Nicoll RA. Tonic activation of NMDA receptors by ambient glutamate enhances excitability of neurons. Science. 1989;246:815–8.
62. Cavelier P, Attwell D. Tonic release of glutamate by a DIDS-sensitive mechanism in rat hippocampal slices. J Physiol. 2005;564:397–410.
63. Herman MA, Jahr CE. Extracellular glutamate concentration in hippocampal slice. J Neurosci. 2007;27:9736–41.
64. Le Meur K, Galante M, Angulo MC, Audinat E. Tonic activation of NMDA receptors by ambient glutamate of non-synaptic origin in the rat hippocampus. J Physiol. 2007;580:373–83.

65. Fleming TM, Scott V, Naskar K, Joe N, Brown CH, Stern JE. State-dependent changes in astrocyte regulation of extrasynaptic NMDA receptor signalling in neurosecretory neurons. J Physiol. 2011;589:3929–41.
66. Povysheva NV, Johnson JW. Tonic NMDA receptor-mediated current in prefrontal cortical pyramidal cells and fast-spiking interneurons. J Neurophysiol. 2012;107:2232–43.
67. Angulo MC, Kozlov AS, Charpak S, Audinat E. Glutamate released from glial cells synchronizes neuronal activity in the hippocampus. J Neurosci. 2004;24:6920–7.
68. Fellin T, Pascual O, Gobbo S, Pozzan T, Haydon PG, Carmignoto G. Neuronal synchrony mediated by astrocytic glutamate through activation of extrasynaptic NMDA receptors. Neuron. 2004;43:729–43.
69. Shigetomi E, Bowser DN, Sofroniew MV, Khakh BS. Two forms of astrocyte calcium excitability have distinct effects on NMDA receptor-mediated slow inward currents in pyramidal neurons. J Neurosci. 2008;28:6659–63.
70. Nie H, Zhang H, Weng HR. Bidirectional neuron-glia interactions triggered by deficiency of glutamate uptake at spinal sensory synapses. J Neurophysiol. 2010;104:713–25.
71. Bergersen LH, Gundersen V. Morphological evidence for vesicular glutamate release from astrocytes. Neuroscience. 2009;158:260–5.
72. Bergersen LH, Morland C, Ormel L, Rinholm JE, Larsson M, Wold JF, Røe AT, Stranna A, Santello M, Bouvier D, et al. Immunogold detection of L-glutamate and D-serine in small synaptic-like microvesicles in adult hippocampal astrocytes. Cereb Cortex. 2012;22:1690–7.
73. Johnson JW, Ascher P. Glycine potentiates the NMDA response in cultured mouse brain neurons. Nature. 1987;325:529–31.
74. Kleckner NW, Dingledine R. Requirement for glycine in activation of NMDA-receptors expressed in Xenopus oocytes. Science. 1988;241:835–7.
75. Hashimoto A, Nishikawa T, Oka T, Takahashi K, Hayashi T. Determination of free amino acid enantiomers in rat brain and serum by high-performance liquid chromatography after derivatization with N-tert.-butyloxycarbonyl-L-cysteine and o-phthaldialdehyde. J Chromatogr. 1992;582:41–8.
76. Hashimoto A, Oka T, Nishikawa T. Anatomical distribution and postnatal changes in endogenous free D-aspartate and D-serine in rat brain and periphery. Eur J Neurosci. 1995;7:1657–63.
77. Hashimoto A, Oka T. Free D-aspartate and D-serine in the mammalian brain and periphery. Prog Neurobiol. 1997;52:325–53.
78. Schell MJ, Molliver ME, Snyder SH. D-serine, an endogenous synaptic modulator: localization to astrocytes and glutamate-stimulated release. Proc Natl Acad Sci U S A. 1995;92:3948–52.
79. Panatier A, Theodosis DT, Mothet JP, Touquet B, Pollegioni L, Poulain DA, Oliet SH. Glia-derived D-serine controls NMDA receptor activity and synaptic memory. Cell. 2006;125:775–84.
80. Papouin T, Oliet SH. Organization, control and function of extrasynaptic NMDA receptors. Philos Trans R Soc Lond B Biol Sci. 2014;369:20130601.
81. Ahmadi S, Muth-Selbach U, Lauterbach A, Lipfert P, Neuhuber WL, Zeilhofer HU. Facilitation of spinal NMDA receptor currents by spillover of synaptically released glycine. Science. 2003;300:2094–7.
82. Kalbaugh TL, Zhang J, Diamond JS. Coagonist release modulates NMDA receptor subtype contributions at synaptic inputs to retinal ganglion cells. J Neurosci. 2009;29:1469–79.
83. Flint AC, Liu X, Kriegstein AR. Nonsynaptic glycine receptor activation during early neocortical development. Neuron. 1998;20:43–53.
84. Ito S, Cherubini E. Strychnine-sensitive glycine responses of neonatal rat hippocampal neurones. J Physiol. 1991;440:67–83.
85. Xu TL, Gong N. Glycine and glycine receptor signaling in hippocampal neurons: diversity, function and regulation. Prog Neurobiol. 2010;91:349–61.

86. Chattipakorn SC, McMahon LL. Pharmacological characterization of glycine-gated chloride currents recorded in rat hippocampal slices. J Neurophysiol. 2002;87:1515–25.
87. Chattipakorn SC, McMahon LL. Strychnine-sensitive glycine receptors depress hyperexcitability in rat dentate gyrus. J Neurophysiol. 2003;89:1339–42.
88. Horio M, Kohno M, Fujita Y, Ishima T, Inoue R, Mori H, Hashimoto K. Levels of D-serine in the brain and peripheral organs of serine racemase (Srr) knock-out mice. Neurochem Int. 2011;59:853–9.
89. Yamamoto S, Morinobu S, Iwamoto Y, Ueda Y, Takei S, Fujita Y, Yamawaki S. Alterations in the hippocampal glycinergic system in an animal model of posttraumatic stress disorder. J Psychiatr Res. 2010;44:1069–74.
90. Holecek M, Kovarik M. Alterations in protein metabolism and amino acid concentrations in rats fed by a high-protein (casein-enriched) diet—effect of starvation. Food Chem Toxicol. 2011;49:3336–42.
91. Okamoto M, Akanuma S, Tachikawa M, Hosoya K. Characteristics of glycine transport across the inner blood-retinal barrier. Neurochem Int. 2009;55:789–95.
92. Zafra F, Aragón C, Olivares L, Danbolt NC, Giménez C, Storm-Mathisen J. Glycine transporters are differentially expressed among CNS cells. J Neurosci. 1995;15:3952–69.
93. Verleysdonk S, Martin H, Willker W, Leibfritz D, Hamprecht B. Rapid uptake and degradation of glycine by astroglial cells in culture: synthesis and release of serine and lactate. Glia. 1999;27:239–48.
94. Song W, Chattipakorn SC, McMahon LL. Glycine-gated chloride channels depress synaptic transmission in rat hippocampus. J Neurophysiol. 2006;95:2366–79.
95. Jonas P, Bischofberger J, Sandkühler J. Corelease of two fast neurotransmitters at a central synapse. Science. 1998;281:419–24.
96. O'Brien JA, Berger AJ. Cotransmission of GABA and glycine to brain stem motoneurons. J Neurophysiol. 1999;82:1638–41.
97. Gielen M, Siegler Retchless B, Mony L, Johnson JW, Paoletti P. Mechanism of differential control of NMDA receptor activity by NR2 subunits. Nature. 2009;459:703–7.
98. Vergnano AM, Rebola N, Savtchenko LP, Pinheiro PS, Casado M, Kieffer BL, Rusakov DA, Mulle C, Paoletti P. Zinc dynamics and action at excitatory synapses. Neuron. 2014;82:1101–14.
99. Paoletti P, Neyton J. NMDA receptor subunits: function and pharmacology. Curr Opin Pharmacol. 2007;7:39–47.
100. Neyton J, Paoletti P. Relating NMDA receptor function to receptor subunit composition: limitations of the pharmacological approach. J Neurosci. 2006;26:1331–3.
101. Isaacson JS, Murphy GJ. Glutamate-mediated extrasynaptic inhibition: direct coupling of NMDA receptors to Ca(2+)-activated K+ channels. Neuron. 2001;31:1027–34.
102. Oikonomou KD, Short SM, Rich MT, Antic SD. Extrasynaptic glutamate receptor activation as cellular bases for dynamic range compression in pyramidal neurons. Front Physiol. 2012;3:334.
103. Liu L, Wong TP, Pozza MF, Lingenhoehl K, Wang Y, Sheng M, Auberson YP, Wang YT. Role of NMDA receptor subtypes in governing the direction of hippocampal synaptic plasticity. Science. 2004;304:1021–4.
104. Barria A, Malinow R. NMDA receptor subunit composition controls synaptic plasticity by regulating binding to CaMKII. Neuron. 2005;48:289–301.
105. Duffy S, Labrie V, Roder JC. D-serine augments NMDA-NR2B receptor-dependent hippocampal long-term depression and spatial reversal learning. Neuropsychopharmacology. 2008;33:1004–18.
106. Morishita W, Lu W, Smith GB, Nicoll RA, Bear MF, Malenka RC. Activation of NR2B-containing NMDA receptors is not required for NMDA receptor-dependent long-term depression. Neuropharmacology. 2007;52:71–6.
107. Zhao JP, Constantine-Paton M. NR2A-/- mice lack long-term potentiation but retain NMDA receptor and L-type Ca2+ channel-dependent long-term depression in the juvenile superior colliculus. J Neurosci. 2007;27:13649–54.

108. Massey PV, Johnson BE, Moult PR, Auberson YP, Brown MW, Molnar E, Collingridge GL, Bashir ZI. Differential roles of NR2A and NR2B-containing NMDA receptors in cortical long-term potentiation and long-term depression. J Neurosci. 2004;24:7821–8.
109. Liu Y, Wong TP, Aarts M, Rooyakkers A, Liu L, Lai TW, Wu DC, Lu J, Tymianski M, Craig AM, Wang YT. NMDA receptor subunits have differential roles in mediating excitotoxic neuronal death both in vitro and in vivo. J Neurosci. 2007;27:2846–57.
110. Xu J, Kurup P, Zhang Y, Goebel-Goody SM, Wu PH, Hawasli AH, Baum ML, Bibb JA, Lombroso PJ. Extrasynaptic NMDA receptors couple preferentially to excitotoxicity via calpain-mediated cleavage of STEP. J Neurosci. 2009;29:9330–43.
111. Olney JW. Brain lesions, obesity, and other disturbances in mice treated with monosodium glutamate. Science. 1969;164:719–21.
112. Lucas D, Newhouse JP. The toxic effect of sodium L-glutamate on the inner layers of the retina. AMA Arch Ophthalmol. 1957;58:193–201.
113. Choi DW, Koh JY, Peters S. Pharmacology of glutamate neurotoxicity in cortical cell culture:attenuation by NMDA antagonists. J Neurosci. 1988;8:185–96.
114. Choi DW, Maulucci-Gedde M, Kriegstein AR. Glutamate neurotoxicity in cortical cellculture. J Neurosci. 1987;7:357–68.
115. Choi DW. Ionic dependence of glutamate neurotoxicity. J Neurosci. 1987;7:369–79.
116. Wroge CM, Hogins J, Eisenman L, Mennerick S. Synaptic NMDA receptors mediate hypoxic excitotoxic death. J Neurosci. 2012;32:6732–42.
117. Talantova M, Sanz-Blasco S, Zhang X, Xia P, Akhtar MW, Okamoto S, Dziewczapolski G, Nakamura T, Cao G, Pratt AE, et al. Aβ induces astrocytic glutamate release, extrasynaptic NMDA receptor activation, and synaptic loss. Proc Natl Acad Sci U S A. 2013;110:2518–27.
118. Zhou X, Ding Q, Chen Z, Yun H, Wang H. Involvement of the GluN2A and GluN2B subunits in synaptic and extrasynaptic N-methyl-D-aspartate receptor function and neuronal excitotoxicity. J Biol Chem. 2013;288:24151–9.
119. Zhou X, Hollern D, Liao J, Andrechek E, Wang H. NMDA receptor-mediated excitotoxicity depends on the coactivation of synaptic and extrasynaptic receptors. Cell Death Dis. 2013;4:560.
120. Katsuki H, Nonaka M, Shirakawa H, Kume T, Akaike A. Endogenous D-serine is involved in induction of neuronal death by N-methyl-D-aspartate and simulated ischemia in rat cerebrocortical slices. J Pharmacol Exp Ther. 2004;311:836–44.
121. Sattler R, Xiong Z, Lu WY, MacDonald JF, Tymianski M. Distinct roles of synaptic and extrasynaptic NMDA receptors in excitotoxicity. J Neurosci. 2000;20:22–33.
122. Amadoro G, Ciotti MT, Costanzi M, Cestari V, Calissano P, Canu N. NMDA receptor mediates tau-induced neurotoxicity by calpain and ERK/MAPK activation. Proc Natl Acad Sci U S A. 2006;103:2892–7.
123. Shleper M, Kartvelishvily E, Wolosker H. D-serine is the dominant endogenous coagonist for NMDA receptor neurotoxicity in organotypic hippocampal slices. J Neurosci. 2005;25:9413–7.
124. Hardingham GE, Fukunaga Y, Bading H. Extrasynaptic NMDARs oppose synaptic NMDARs by triggering CREB shut-off and cell death pathways. Nat Neurosci. 2002;5:405–14.
125. Emnett CM, Eisenman LN, Taylor AM, Izumi Y, Zorumski CF, Mennerick S. Indistinguishable synaptic pharmacodynamics of the N-methyl-D-aspartate receptor channel blockers memantine and ketamine. Mol Pharmacol. 2013;84:935–47.
126. Xia P, Chen HS, Zhang D, Lipton SA. Memantine preferentially blocks extrasynaptic over synaptic NMDA receptor currents in hippocampal autapses. J Neurosci. 2010;30:11246–50.
127. Ditzler K. Efficacy and tolerability of memantine in patients with dementia syndrome. A double-blind, placebo controlled trial. Arzneimittelforschung. 1991;41:773–80.
128. Lipton SA. The molecular basis of memantine action in Alzheimer's disease and other neurologic disorders: low-affinity, uncompetitive antagonism. Curr Alzheimer Res. 2005;2:155–65.
129. Pál B. Astrocytic actions on extrasynaptic neuronal currents. Front Cell Neurosci. 2015;9:474.

130. Wu YW, Grebenyuk S, McHugh TJ, Rusakov DA, Semyanov A. Backpropagating action potentials enable detection of extrasynaptic glutamate by NMDA receptors. Cell Rep. 2012;1:495–505.
131. Riebe I, Seth H, Culley G, Dósa Z, Radi S, Strand K, Fröjd V, Hanse E. Tonically active NMDA receptors—a signalling mechanism critical for interneuronal excitability in the CA1 stratum radiatum. Eur J Neurosci. 2016;43:169–78.
132. Ortinski PI, Turner JR, Pierce RC. Extrasynaptic targeting of NMDA receptors following D1 dopamine receptor activation and cocaine self-administration. J Neurosci. 2013;33:9451–61.

Chapter 3
Functional Distribution and Regulation of the NMDAR in the Kidney, Heart and Parathyroid Gland

Milica Bozic and José M. Valdivielso

Abstract *N*-Methyl-D-aspartate receptor (NMDAR) is an ionotropic glutamate receptor with a high permeability to calcium and a unique feature of controlling numerous calcium-dependent processes. Apart of being widely distributed in the central nervous system, presence of NMDAR and its potential significance in a variety of non-neuronal cells and tissues has become an interesting research topic. This chapter summarizes prevailing knowledge on the functional distribution and regulation of NMDARs in the kidney and parathyroid gland, the two organs important in calcium homeostasis, as well as in the heart, the organ whose function is highly dependable on balanced intracellular calcium concentrations. The chapter also examines studies that have advanced our understanding on the multiple roles of NMDAR in different physiological and pathological processes in the kidney, heart and parathyroid gland.

Keywords Kidney • Heart • Parathyroid gland • Proximal tubule • Cardiomyocytes • Podocyte • NMDAR • Calcium

Abbreviations

$1,25(OH)_2D_3$	1,25-Dihydroxyvitamin D
2HPT	Secondary hyperparathyroidism
AKI	Acute kidney injury
Akt	Protein kinase B (PKB)
ALP	Alkaline phosphatase
CKD	Chronic kidney disease

M. Bozic (✉) • J.M. Valdivielso
Nephrology Department, Institute for Biomedical Research (IRB Lleida), Edificio Biomedicina 1. Lab B1-10, Av. Rovira Roure 80, 25198 Lleida, Spain
e-mail: bozicm@medicina.udl.cat; valdivielso@medicina.udl.cat

K. Hashimoto (ed.), *The NMDA Receptors*, The Receptors 30,
DOI 10.1007/978-3-319-49795-2_3

EAATs	Glutamate transporters (excitatory amino-acid transporters)
EMT	Epithelial-mesenchymal transition
Erk	Extracellular receptor kinase
FGF23	Fibroblast growth factor 23
GFR	Glomerular filtration rate
Hcy	Homocysteine
hHcys	Hyperhomocyteinemia
HIV-1	Human immunodeficiency virus type I
HK-2	Human proximal tubular cells
HRV	Heart rate variability
iGluRs	Ionotropic glutamate receptors
IMCDs	Inner medullary collecting duct cells
IRI	Ischemia reperfusion injury
LLC-PK1	Pig kidney epithelial cells
MDCKs	Madin-Darby canine kidney cells
MEK	MAPK/Erk kinase
MK-801	(+)-5-methyl-10,11-dihydro-5H-dibenzo(a,d)cyclohepten-5,10-imine maleate
MMP-9	Matrix metallopeptidase 9
NMDA	*N*-Methyl-D-aspartic acid
NMDAR	*N*-Methyl-D-aspartate receptor
nNOS	Neuronal nitric oxide synthase
NO	Nitric oxide
PKA	Protein kinase A
PTG	Parathyroid gland
PTH	Parathyroid hormone
Ras	Rat Sarcoma protein (Ras) subfamily of small GTPases
RhoA	Family member of the Rho-like GTPases
ROS	Reactive oxygen species
SNGFR	Single nephron glomerular filtration rate
TGF	Tubuloglomerular feedback
TGF-β1	Transforming growth factor beta-1
α-SMA	Alpha smooth muscle actin

3.1 Distribution of NMDAR in Peripheral Tissues

In addition to NMDAR's broad distribution in neurons, early work from Nishikawa [1] and Moroni [2] indicated that NMDA receptors could be present in extraneuronal tissues. More recent reports from several laboratories around the globe have demonstrated that functional NMDARs are also expressed in a variety of non-neuronal cells and tissues such as human keratinocytes [3, 4], lymphocytes [5], artery [6–8], bone cells [9, 10], embryonic [11] and adult heart [7, 8, 12, 13], rat cardiocytes [3], lung, thymus, stomach [8], parathyroid gland [14], ovaries, spleen [15], skeletal muscle, pancreas [16], lower urogenital tract [17], renal pelvis [18] and kidney [8, 12, 19, 20] (Table 3.1).

According to the study of differential binding of NMDAR's antagonists, Nasstrom et al. [21] suggested the presence of this receptor in various tissues outside the CNS, such as heart, stomach, pancreas, and kidney. For instance, GluN1 subunit was found in the adult rat heart [7, 8] and heart cells grown in culture [8]. The GluN2 protein expression was not detected in the adult heart [8], while in the newborn heart it was found only transient GluN2B mRNA and protein expression [11]. Chen et al. [6] confirmed the presence of all NMDAR subunits in the rat aortic endothelial cells, as well as the GluN1 [6, 8] and GluN2A in rat carotid artery [6]. Results from Deng et al. [22] showed the presence of GluN1 on the basolateral side of the proximal tubules of the kidney. Leung et al. [8] confirmed the presence of GluN1 in rat kidney cortex and medulla, while of the other NMDAR subunits, only GluN2C was detectable in the rat kidney. Both GluN1 and GluN2C are present in MDCK cells, opossum kidney and LLC-PK1 cells [8], whereas HK-2 human kid-

Table 3.1 Overview of the tissue expression of NMDAR subunits in peripheral tissues

NMDAR subunit	Organ	Tissue/cell type	Reference
GluN1	Kidney	Kidney (cortex, medulla), glomeruli, tubules, podocytes, HK-2, MDCKs, OKs, LLC-PK1, IMCDs	[8, 19, 20, 22–26]
	Heart	Heart, heart myoblast cells, cardiocytes, carotid artery, RAECs	[3, 6–8, 13, 15]
	Parathyroid gland		[14]
	Others	Lymphocytes, keratinocytes, spleen, adrenal gland, ovaries, stomach, lower urogenital tract, renal pelvis, thymus, bone	[3–5, 8, 9, 15, 17, 18, 27–29, 31, 88]
GluN2A	Kidney	Glomeruli, HK-2	[19, 23]
	Heart	Carotid artery, RAECs	[6]
GluN2B	Kidney	Kidney cortex, HK-2	[19, 20]
	Heart	Newborn heart, RAECs	[6, 7]
	Others	Lymphocytes, osteoclasts	[5, 28]
GluN2C	Kidney	Kidney (cortex, medulla), HK-2, MDCKs, OKs, LLC-PK1, IMCDs	[8, 19, 20, 26]
	Heart	RAECs	[6]
	Others	Pancreas, skeletal muscle	[16]
GluN2D	Kidney	Kidney cortex, HK-2	[19, 20]
	Heart	RAECs	[6]
GluN3A	Kidney	Kidney, IMCDs	[26]
GluN3B	Kidney	Kidney, IMCDs	[26]

RAECs, rat aortic endothelial cells; HK-2, human proximal tubular cells; MDCKs, Madin-Darby canine kidney cells; OKs, opossum kidney cells; LLC-PK1, pig kidney epithelial cells; IMCDs, inner medullary collecting duct cells

ney cells express GluN1 and all four GluN2 subunits [19]. Zhang et al. [23] demonstrated distribution of GluN1 and GluN2A in the rat glomeruli. Furthermore, functional NMDAR was found present in human [24] and mouse [24, 25] podocytes. Sproul et al. [26] showed high protein expression of NMDAR subunits GluN3A and GluN3B in the neonatal kidney and suggested that there was continued expression of GluN3A protein in the renal medulla and papilla of the adult mouse.

It has also been reported that GluN1 mRNA and protein is broadly expressed in osteoblasts and osteoclasts from different species [9, 27–31]. Nevertheless, the pattern of GluN2 expression changes with different species and/or with the differentiation state of the cells. Thus, Itzstein et al. [28] demonstrated presence of GluN2B and GluN2D in human and rat osteoclasts, while osteoclastic progenitor cells express only GluN2A mRNA [32]. In primary rat calvarian osteoblasts, the exclusive expression of GluN2D mRNA has been shown [9, 33]. However, reports from Itzstein et al. [28], Kalariti et al. [31] and Gu et al. [34] showed contradictory results. A recent report by Fujita et al. [35] has also demonstrated the presence of NMDAR in osteocytes. The role of NMDAR in the bone will be deeply covered in another section of this book. Thus, in the present chapter we will focus on the role of NMDAR in the kidney, parathyroid gland and the heart.

3.2 Functional Distribution of NMDAR in the Kidney

Kidneys are essential excretory organs of the body responsible for preserving the internal environment of the organism. Glomerular filtration, tubular reabsorption, and tubular excretion are three mechanisms by which kidneys accomplish the homeostasis of the internal environment. The kidneys display remarkable ability of blood flow autoregulation and a variety of overlapping neurohormonal factors that unify to regulate kidney blood flow and glomerular ultrafiltration [36].

Importance of the NMDAR in the kidney and its functional role has emerged as an interesting research topic in the past decade, although experimental data are scarce. Deng et al. [12] were among first to show the presence of GluN1 subunit of NMDAR in the basolateral proximal tubules of the rat kidney and confirmed a role for renal NMDAR in the regulation of renal vasodilation. To study the functional role of NMDAR in the kidney, renal hemodynamic effects of NMDAR inhibition were assessed in Wistar rats using an antagonist of the NMDAR (MK-801) or an inhibitor of glycine binding to NMDAR (5,7-dichlorokynurenic acid). Both antagonists caused renal vasoconstriction and attenuated subsequent renal vasodilatory response to glycine infusion, effects not mediated by renal inervation [12]. These results have been confirmed by studies of Bądzyńska et al. [37] in which the administration of glycine increased renal blood flow (RBF) and cortical renal plasma flow (CRPF) both in normal rats and in spontaneously hypertensive rats (SHR), without systemic changes in blood pressure. In addition, administration of glycine induced diuresis and natriuresis, although the effect was less effective in SHR, suggesting a decrease in NMDAR in this model of hypertension. Seeking to elucidate the mechanism responsible for the initiation of renal vasodilatatory response to glycine infusion, Slomowitz et al. [38] demonstrated that low protein feeding in rats resulted in a loss of glycine-induced vasodilatation and glomerular filtration rate (GFR) response which was associated with a significant decrease in renal NMDAR protein expression. The results of this study stressed the potential role of NMDAR in modulation of proximal tubular reabsorption and GFR response. According to Deng et al. [22] renal NMDARs independently stimulate proximal tubular reabsorption and glomerular filtration, a fact consistent with the presence of NMDARs in both, glomerular and tubular cells. In the mentioned study, systemic administration of MK-801, as well as direct application to the glomerulus or proximal tubule by microperfusion, caused significant reduction of single nephron glomerular filtration rate (SNGFR) and lessened proximal tubular reabsorption in kidneys of Wistar rats. Furthermore, NMDAR blockade suppressed proximal reabsorption independent of the filtered load and reduced SNGFR independent of tubuloglomerular feedback (TGF). The authors concluded that NMDARs in the hydropenic rat kidney cortex tonically affected vasodilation and provided a stimulus for proximal tubular reabsorption [22]. The results discussed above suggest that NMDAR modulators that exert a tonic vasodilatatory effect on the glomerular microvasculature might be used as a useful therapeutic tool to regulate TGF and glomerular filtration.

These previous results also suggest that excessive consumption of diet-associated glutamate can have a potential effect on renal function. Indeed, monosodium glutamate (MSG) consumption has been steadily increasing worldwide in recent years as flavoring in cooking to increase palatability and food selection in a meal [39]. A very recent report of Mahieu et al. [40] demonstrated that the addition of MSG in the diet of rats increased both GFR and CRPF with an increase of absolute sodium reabsorption. Those results were parallel to an increase in NMDAR immunostaining in the kidney, suggesting that activation of NMDAR induces hyperfiltration with secondary increased tubular reabsorption of Na. Furthermore, MSG also produces up-regulation of GluN1 in kidneys. Inhibition of NMDAR with MK-801 significantly reduced both the GFR and CRPF, although the percentage of reduction was higher in the animals supplemented with MSG. MK-801 also reduced fractional excretion of water, sodium and potassium. Therefore, chronic activation of NMDAR over time may cause alterations at both renal and systemic level, such as renal failure and hypertension.

Most glomerular diseases are characterized by defect of the filtration barrier, where podocytes play a central role. Giardino et al. [25] report that NMDAR plays an important role in maintaining the stability of the glomerular filtration barrier, while derangements in glutamatergic signaling in podocytes increase albuminuria leading to proteinuric kidney disease. Namely, in the podocyte culture, treatment with antagonists of the NMDAR, norketamine hydrochloride or MK-801, caused a profound remodeling of podocyte cytoskeleton and disappearance of nephrin from podocyte processes, while the addition of the agonist NMDA (50 μM; 15 min) reversed these changes [25]. The authors demonstrated that both antagonists increased significantly glomerular albumin permeability in isolated rat glomeruli proving that the NMDAR blockade had a direct effect on glomerular filtration. In vivo, Balb/c mice treated with norketamine for 3 days showed a marked increase in U_{Alb}/U_{Creat} followed by a decreased expression of glomerular nephrin in treated animals [25]. Recent report from Anderson et al. [24] showed that sustained exposure to 50 μM NMDA caused a reduction in the expression of nephrin in podocyte cell lines. In addition, the same authors demonstrated that sustained activation of NMDARs activate secondary signaling cascades known to be important for podocyte function, such as phosphorylation of Erk1/2 and Akt, as well as the activation of RhoA [24]. The authors concluded that hyperactivation of NMDAR could be deleterious for the process of glomerular filtration, inducing the loss of proteins essential for the normal function of slit diaphragms. The same group demonstrated that treatment of podocytes with NMDA for 24 h reduced total and cell surface expression of essential podocyte markers nephrin and podocin [41]. Additional exposure of podocytes to NMDA, in this study, for a period of 72 h evoked a significant apoptotic cell death [41]. In accordance with the above mentioned, Zhang et al. [23] pointed out to a role of NMDAR in hyperhomocyteinemia (hHcys)-induced glomerulosclerosis. They reported that both GluN1 and GluN2 subunits were increased in the glomeruli of rats with hHcys, which was inhibited by treatment with MK-801, indicating the involvement of this receptor in the pathogenesis of hHcys-induced glomerulosclerosis [23].

Recent data also imply to a role of NMDAR on the onset of diabetic nephropathy, namely in extensive extracellular matrix (ECM) remodeling [42]. Thus, in diabetic mice, GluN1 is induced, and this increase in glutamatergic signaling also increases conexin-40 and -43 expressions which are involved in ECM remodeling. The upregulation of NMDAR seems to be downstream of a rective oxygen species (ROS)-induced increase of MMP-9 levels and a decrease of hydrogen sulfide (HS) in diabetic kidney, which indicates an active remodeling process due to increased oxidative radicals as suggested by increased levels of N-tyrosine. Those results were further confirmed by a follow up paper in which treatment of the animals with HS decreased the NMDAR overexpression in diabetic kidneys and restored renal function and excessive remodeling [43]. Those results are in agreement with a report of Szaroma et al. [44] which showed that activation of NMDAR in the kidney decreased antioxidant capacity by decreasing the activity of superoxide dismutase, catalase and glutathione peroxidase and the amount of reduced glutathione.

The important reno-protective role of GluN3A subunit of the NMDAR in inner medullary collecting duct (IMCD) cells was demonstrated by Sproul et al. [26]. Specifically, this group demonstrated that the knockdown of GluN3A in IMCD cells led to an increase of basal intracellular calcium concentration, reduced cell growth, higher rate of cell death and reduced water transport in response to the addition of vasopressin. The authors concluded that GluN3A subunit may have a protective role in IMCDs through regulation of the intracellular calcium levels which enables the principal cells of the collecting duct to reabsorb water and thereby increase medullary osmolality [26]. Thus, GluN3A may be an important target for protection of IMCDs and an important mechanism that protects the function of the principal cells to reabsorb water, thereby helping to maintain the countercurrent multiplication system [45, 46].

Studies by Leung et al. [47, 48] using in vivo and in vitro approach, investigated the possible role of NMDAR in renal cell toxicity. Taking into account that NMDAR plays a pivotal role in gentamicin ototoxicity [49–51], the fact that is expressed in the renal proximal tubule [47], as well as the high degree of gentamicin nephrotoxicity reported previously [47], the authors speculated that the NMDAR might be activated by gentamicin and may play a role in renal injury caused by this antibiotic. The experiments conducted on Sprague-Dawley rats showed that GluN1 and GluN2C subunits were substantially increased in the renal cortex of short-term gentamicin rats and the receptor likely mediated cell damage via the endothelin-ETBP-nitric oxide pathway [47]. Described renal damage in rats exposed to short-term gentamicin was attenuated after exposure to MK-801, proving the important role of NMDAR in the gentamicin model of renal toxicity. Leung et al. [48] in their in vitro approach using MDCK cells and proximal tubule-like opossum kidney cells, further demonstrated that excessive stimulation of the NMDAR with 10 mM glutamate, as well as the excessive blockade of this receptor with MK-801 or CPP, resulted in deleterious effects on cell survival [48].

Acute kidney injury (AKI) is a clinical syndrome characterized by rapid decrease in renal function and is one of the major health problems worldwide [52]. One of the principal causes of AKI is renal ischemia reperfusion injury (IRI) that is associated with reduced oxygen and nutrient supply leading to apoptotic and necrotic death of tubular cells, and subsequent impairmet of renal function [53, 54]. Using unilateral ischemia reperfusion (IR) in rats, Yang et al. [53] showed that IR in kidneys enhanced renal GluN1 protein expression and was associated with decreased GFR response. Intrarenal arterial NMDA infusion decreased GFR in the control and ischemia-reperfusion kidneys. NMDAR antagonist, AP-5, was able to completely abolish NMDA-induced renal dysfunction and to ameliorate IR-induced glomerular and tubular dysfunction in both groups of animals. Pundir et al. [54] confirmed that AKI is associated with the activation of NMDARs and oxidative stress. The authors demonstrated that various NMDAR antagonists including kynurenic acid, ketamine and channel blocking agent magnesium sulphate attenuated ischemia-reperfusion-induced AKI and reduced oxidative stress, suggesting the beneficial effect of the antagonism of various allosteric sites of NMDAR against IR-induced AKI [54]. Recently published study of the same group suggests that glycine increases ischemia reperfusion-induced AKI through NMDAR activation in rats rather than strychnine-sensitive glycinergic receptors [52]. In a very recent paper, Shing et al. also confirmed those results and demonstrated that Pioglitazone, a PPAR-c agonist that attenuates ischemia-reperfusion-induced renal damage, exert its protective function by inhibiting NMDAR as Pioglitazone protective effect is attenuated by previous activation of NMDAR [55]. Another piece of evidence linking excessive NMDAR signaling with AKI has been provided by Lin et al. [56]. Thus, in rat model of lipopolysaccharide (LPS)-induced endotoxemia, an increase of GluN1 was found in renal tubules. LPS-induced renal damage was ameliorated by the NMDA receptor blocker, MK-801. LPS also induced cell damage in cultured tubular cell lines and primary rat proximal tubular cells, which was also mitigated by MK-801 and by small interfering RNA targeting GluN1. In this case, the increase in NMDAR seems to be mediated by an increase in Il1β. Furthermore, a paper of Cauli et al. shows that in rats with acute liver failure (a lethal condition that leads to rapid progression of deleterious events including renal failure), antagonism of NMDAR delays death by increasing GFR and, therefore, the clearance of ammonia which is cytotoxic in the brain [57]. Thus, it seems clear that in AKI, an excessive glutamatergic signaling through NMDAR is deleterious to the kidney, and NMDAR antagonism could be a therapeutic option to improve renal function.

Results from our group pointed to an indispensable role of NMDAR in the preservation of normal epithelial phenotype of proximal tubular cells and in the modulation of important steps of tubular epithelial-mesenchymal transition (EMT) [20]. Thus, knockdown of GluN1 induced remarkable changes in epithelial phenotype of human proximal tubular (HK-2) cells, manifested as a decrease of E-cadherin and an increase of α-SMA, along with the changes in cell morphology [20]. Furthermore, in vitro, TGF-β1-induced EMT in HK-2 cells was attenuated by co-treatment with NMDA. The mechanism behind the effect of NMDA on TGF-β1-induced tubular EMT could be related to the inhibition of the Ras-MEK pathway [20]. In an in vivo study, administration of NMDA significantly inhibited the expression of markers of renal fibrosis in the obstructed mouse kidneys, pointing to a paramount role of NMDAR in the preservation of normal epithelial phenotype of proximal tubular cells and in the modulation of important steps of tubular EMT [20].

Another line of research from our group addressed the role of NMDAR in the regulation of the renal active vitamin D synthesis. Indeed, we demonstrated that the activation of NMDAR caused a decrease in the synthesis of 1,25 $(OH)_2D_3$ in proximal tubular cells in vitro and of $1,25(OH)_2D_3$ levels in the blood in vivo [19]. This effect, inhibiting active vitamin D synthesis, was due to a decrease in the levels of 1α-hydroxylase, mediated by an activation of the MAPK pathway [19]. Therefore, lack of the inhibitory effect of vitamin D on the parathyroid gland resulted in an increase of parathyroid hormone (PTH) synthesis and release. Furthermore, animals with induced chronic kidney disease (CKD) demonstrated high levels of renal glutamate compared with healthy ones, pointing to overactivation of tubular NMDAR by glutamate as a possible cause for the downregulation of 1α-hydroxylase, subsequent drop in $1,25(OH)_2D_3$ synthesis and the onset of secondary hyperparathyroidism (2HPT) associated with CKD.

Thus it seems clear that basal activation of NMDAR in the kidney is indispensable for the maintenance of normal tubular and kidney function, while an excessive activation can cause a disturbance of kidney homeostasis leading to a variety of pathophysiological consequences.

3.3 Functional Distribution of NMDAR in the Heart

It has been suggested that activation of the NMDAR in the heart has an important impact on electrical activity of this organ and may play a role in cardiac arrhythmogenesis [58]. The recent findings from Shi et al. [59] show that the chronic NMDA administration induced significant cardiac electrophysiological alterations and increased susceptibility to ventricular arrhythmias (VAs), together with mild myocardial interstitial fibrosis. Liu et al. [60] demonstrated that the activation of NMDAR induced ventricular tachycardia and ventricular fibrillation in rats with myocardial infarction, but not in normal rats. Consistent with these results, Maldonado et al. [61] demonstrated that overactivation of the cardiac NMDAR with high levels of homocysteine (Hcy) or other circulating excitatory amino acids, led to an increase of intracellular calcium which may alter the rate of cell-to-cell conduction and create arrhythmogenic substrate. These findings suggest that in the setting of myocardial ischemia, blocking the NMDAR is important in reducing arrhythmogenic substrate. Thus, D'Amico et al. [58] demonstrated that blockade of the NMDAR reduced reperfusion-induced arrhythmias, but not ischemia-induced arrhythmias. Furthermore, recent findings by Sun et al. [62] demonstrate that the inhibition of NMDAR significantly ameliorates ischemia-reperfusion-induced ventricular arrhythmias in rats, and reduces calcium accumulation in mitochondria. To define the mechanism of NMDAR blockade on cardiac action, Huang et al. [63] examined the effects of MK-801 on heart rate and contractility of isolated rat cardiac preparations in vitro. Moreover, this study assessed the effects of MK-801 on action potential and membrane ionic currents of rat ventricular myocytes. Their results revealed that the inhibition of NMDAR caused bradycardia and increased cardiac contraction in rat ventricular myocytes. Nevertheless, neither the twitch tension nor the heart rate was significantly affected by NMDA, implying that bradycardia and the increase of cardiac contractility cannot be attributed to the blockade of NMDARs in cardiac tissues. Indeed, the authors show that MK-801 prolonged cardiac action potential duration via inhibiting potassium outward current. Reduced heart rate variability (HRV) has been shown to be a significant predictor of mortality after myocardial infarction [64]. In the study of Bennett et al. [65], NMDAR blockade provoked an increase of HRV in adolescent rhesus monkeys. In rodents, antagonizing NMDAR by ketamine increased HRV and had anti-arrhythmic effects [58, 66, 67]. Recent results from Shi et al. [68] demonstrated that NMDAR activation reduced HRV and increased susceptibility to atrial fibrillation in rats, with cardiac autonomic dysfunction, atrial fibrosis, and loss of gap junction identified as potential mechanistic contributors [68]. Furthermore, Hageman et al. [69] showed that inhibition of NMDAR significantly decreased pacing-induced ventricular arrhythmias in dogs exposed to cocaine.

Results from McGee et al. [7, 8] confirm the presence of GluN1 in the vasculature and heart and provide evidence of an important role of NMDAR in blood pressure regulation. In their study, NMDA administration elicited dose-related pressor response and an increase in heart rate, which was not dose-dependent. The NMDA-evoked pressor and tachycardiac responses were mediated via the activation of vascular NMDAR because it persisted after ganglion blockade (hexamethonium) and was attenuated by the NMDAR antagonists. Furthermore, NMDA increased the contractile response of isolated aortic rings.

Results from Gao et al. [70] suggested the important role for NMDAR in myocardial pathogenesis. Namely, these authors showed that activation of NMDAR in rat cardiomyocytes led to an increased oxidative stress and calcium load in mitochondria, inducing apoptosis. This cytotoxic effect was abolished by NMDAR blockade and by free radical scavengers [70]. Srejovic et al. [71] demonstrated that the application of homocysteine thiolactone and MK-801 induced significant decrease in oxidative stress parameters. Recent work from Tyagi et al. [72] provides evidence about the involvement of NMDAR in autophagy. Cardiomyocyte autophagy is an important process in the pathogenesis of cardiovascular diseases. Thus, a study of Tyagi et al. [72] revealed that cardiomyocyte-specific deletion of GluN1 led to a reduction of Hcy-induced myocyte mitochondrial ROS, NO and MMP-9 levels in cardiac mitochondria, and subsequently amelioration of mitophagy. Moshal et al. [73] showed that cardiomyocyte specific deletion of GluN1 attenuated Hcy-induced increase in GluN1 protein expression and Hcy-induced mitochondrial permeability transition. The same research group showed that the blockade of NMDAR, by MK-801, attenuated Hcy-induced membrane permeability transition in cardiomyocytes [73]. Recent results from Srejovic et al. [71] demonstrated that administration of homocysteine thiolactone, as well as MK-801 alone, induced a decrease in cardiac contractility, systolic pressure in left ventricle (SLVP), heart rate and coronary flow relative to the control group. The authors point out that negative effects exerted by Hcy on NMDA receptor activity are not mediated by oxidative stress [71]. Nevertheless, the exact pathway of how the NMDAR activation alters the myocyte physiology in the setting of homocysteinemia remains to be determined [74]. Consistent with previous results, Meng et al. [75] provide another evidence about the involvement of NMDAR in autophagy. In their study, inhibition of NMDAR in cardiomyocytes blocked the increase of autophagic proteins and autophagosomes, and subsequent autophagy induced by HIV-1 gp120. Another evidence, supporting the idea that NMDAR blockade may act as a cardioprotective strategy, comes from Meneghini et al. [76] who showed that the effective blockade of NMDAR by memantine prevented nuclear size reduction of cardiomyocytes in the left ventricles of animals exposed to cold stress.

In conclusion, NMDAR has a paramount role in the pathophysiology of the cardiovascular system. Activation of NMDAR promotes ventricular arrhythmias, oxidative stress and autophagy of cardiomyocytes, while antagonizing this receptor increases HRV, has an anti-arrhythmic effect and decreases incidence of myocardial pathologies. These findings may provide further insight into a novel therapeutic target for heart disease associated with NMDAR activation.

3.4 Functional Distribution of NMDAR in the Parathyroid Gland

The presence of NMDARs in the three main organs regulating mineral metabolism (bone, PTG and kidney) points to an important role of glutamate in regulating calcium and phosphorus levels. Thus, the activation of NMDAR in the PTG induced a decrease of PTH synthesis and secretion. This direct effect was demonstrated in vitro and in vivo 60 min after NMDA administration. Furthermore, administration of NMDA blunted the increase of PTH induced in vivo by hypocalcemia, but not in animals with 2HPT [14]. However, sustained activation of NMDAR in vivo has shown to induce an increase of PTH and of PTH mediated bone remodeling markers [19], which could be mediated by both, the decrease in PTH and a direct effect of NMDA on bone cells. This increase in PTH seems to be mediated by a decrease in the synthesis of active vitamin D in the tubular cells, because external administration of active vitamin D compounds can block the increase in PTH mediated by repeated administration of NMDA [19]. These results point to a role of glutamate in the onset of 2HPT in CKD. Although a decrease in the circulating levels of active vitamin D seems to be an accepted mechanism, the cause for that decrease has not yet been fully elucidated. An increase of glutamate levels, which have been reported in animals with CKD [19] could be playing a role, together with some other mediators like FGF23 [77]. However, that increase in glutamate should be directly inhibiting the synthesis and release of PTH in the PTG. Nevertheless, as in the case of Klotho (the receptor for FGF23 which also increases in CKD and directly inhibits PTH in the PTG) [77], the levels of the NMDAR decrease in the hypertrophic gland, rendering them insensitive to the inhibitory effect of increased glutamate levels [14].

3.5 Regulation of NMDAR Activation in Peripheral Tissues

Glutamate (Glu) is the most important excitatory neurotransmitter within CNS, alongside glycine. Under normal heathy conditions, glutamate is found in cerebrospinal fluid and brain extracellular fluid, reaching the maximum concentration of 1 μM [78], while in the plasma glutamate concentration ranges between 10–50 μM [79]. Therefore, it seems logical to think that NMDARs in peripheral tissues could be under tonic activation. However, this not seems to be the case. Thus, administration of NMDA or glutamate in vivo seems to increase the activation of NMDARs in kidney and parathyroid gland [7, 19, 20, 22, 24, 25, 41, 60, 61, 70, 80]. This fact could be explained by several reasons. First, it is possible that interstitial levels of glutamate are not accurately reflected by plasma levels. In the CNS, extracellular glutamate concentrations are tightly regulated by the function of glutamate transporters

(EAATs). EAATs have a role in termination of transmitter action by removing released neurotransmitters from the synaptic cleft [81], thus preventing the overstimulation of the postsynaptic glutamate receptors. Nevertheless, EAATs have also been detected in a variety of peripheral tissues such as bone [81, 82], heart [83], kidney [84, 85], etc. Presence of different isoforms of EATTs in peripheral cells and tissues implies that glutamate may play a role in paracrine signaling in peripheral tissues. Thus, and although difficult to prove, the concentrations of glutamate in the extracellular space in target tissue may be regulated by EAATs.

A second possibility is that NMDARs from peripheral tissues have lower affinities for glutamate and glycine than neuronal NMDARs. Namely, it has been reported that glutamate binds to osteoblast's NMDAR, with a Kd of approximately 10^{-4} mol/L (100 μM) [86], while in the brain it reaches only 36 nM. Furthermore, Anderson et al. [24] reported that podocyte's NMDAR, while readily activated by NMDA, do not respond robustly to glutamate, aspartate or glycine, even at high concentrations (10 mM). The authors implied to a possibility that these amino acids do not have an access to the ligand-binding pockets on podocyte GluN2 subunits and propose the role for a local glutamatergic signaling system in suppressing tonic activation of NMDARs by different circulating factors [24].

In addition, plasma levels of glutamate also fluctuate in some conditions. Thus, elevated plasma Glu has been reported in uremic patients on haemodialysis [79] and various types of tumors (reviewed by [87]). Furthermore, elevated levels of Glu have been reported in the synovial fluid obtained from patients with arthritis [82].

3.6 Summary and Conclusions

It is clear from evidence reviewed in this chapter that NMDARs are broadly distributed outside the CNS and have multiple roles in physiological and pathological processes in the kidney, heart and parathyroid gland (Table 3.2). Furthermore, each of these diverse functions of NMDAR holds an important therapeutic potential for the management of different diseases. As highlighted in this chapter, both activation and blockade of NMDAR could have diverse (beneficial/deleterious) effects on the given tissue and/or organ. Thus, activation of NMDAR in the heart promotes ventricular arrhythmias, oxidative stress and autophagy of cardiomyocytes, while antagonizing this receptor increases heart rate variability, has an anti-arrhythmic effect and decreases incidence of myocardial pathologies. Nevertheless, the blockade of NMDAR in the kidney does not always necessarily leads to a beneficial effect. Namely, on the one hand, blockade of the NMDAR in the kidney ameliorates ischemia-reperfusion-induced glomerular and tubular dysfunction. On the other hand, antagonizing NMDAR in podocytes caused a disturbance of the glomerular filtration barrier. Activation of NMDAR in the kidney shows a more consistent effect, increasing renal blood flow and ameliorating tubulointestinal fibrosis in vivo. Thus, the possibility of a beneficial combined effect of inhibition of tubular EMT and vasodilatation should not be neglected in some kidney conditions. Nevertheless,

Table 3.2 Overview of the reported effects of NMDAR activation/blockade on organ function

Organ	Modulator	Mode of action	Effect on organ function	Reference
Kidney	MK-801/ 5,7-dichloro kynurenic acid	Blockade	Renal vasoconstriction	[12]
	MK-801	Blockade	Reduction of SNGFR	[22]
	Norketamine hydrochloride/ MK-801	Blockade	Remodeling of podocyte cytoskeleton; increase of glomerular albumin permeability	[25]
	NMDA	Activation	Reduction of nephrin and podocyn expression; apoptosis	[24, 41]
	NMDA	Activation	Decrease of GFR	[53]
	D-AP5	Blockade	Attenuation of ischemia-reperfusion-induced glomerular and tubular dysfunction	[53]
	Kynurenic acid, ketamine, magnesium sulphate	Blockade	Attenuation of ischemia-reperfusion-induced AKI and oxidative stress	[54]
	NMDA	Activation	Attenuation of TGF-β1-induced EMT; inhibition of renal fibrosis in vivo	[20]
	NMDA	Activation	Decrease in the synthesis of $1,25(OH)_2D_3$ in proximal tubular cells	[19]
Heart	NMDA	Activation	Ventricular tachycardia, arrhythmia	[6, 60]
	MK-801	Blockade	Reduction of reperfusion-induced arrhythmias	[62, 89]
	MK-801, ketamine	Blockade	Bradycardia, increase in HRV	[58, 63, 65–67]
	NMDA	Activation	Increased oxidative stress and apoptosis	[70]
	MK-801	Blockade	Blockade of autophagy	[72]
	Memantine	Blockade	Prevention of nuclear size reduction of cardiomyocytes	[76]
PTG	NMDA	Activation	Decrease of PTH synthesis and secretion	[14]

SNGFR, single nephron glomerular filtration rate; GFR, glomerular filtration rate; AKI, acute kidney injury; EMT, epithelial-mesenchymal transition; PTH, parathyroid hormone; HRV, heart rate variability

excessive activation can cause a disturbance of kidney homeostasis leading to a variety of pathophysiological consequences. As discussed above, hyperactivation of NMDAR could induce apoptotic cell death and decrease active vitamin D synthesis, which could lead to a common complication in CKD patients, the 2HPT. This complication also seems to involve excessive activation of NMDARs in bone and a decrease of NMDAR signaling in the PTG.

Comprehensive studies of the functional divergence of various NMDAR modulators and the development of new ones are required in order to broad our knowledge on working principles of NMDARs in the periphery, as well as to establish powerful clinical interventions for different diseases. Furthermore, it will help us to determine whether NMDARs comprise potential targets in particular organs and whether NMDAR activators or inhibitors would be needed to positively affect disease outcomes.

References

1. Nishikawa T, Morita K, Kinjo K, Tsujimoto A. Stimulation of catecholamine release from isolated adrenal glands by some amino acids. Jpn J Pharmacol. 1982;32:291–7.
2. Moroni F, Luzzi S, Franchi-Micheli S, Zilleti L. The presence of N-methyl-D-aspartate-type receptors for glutamic acid in the guinea pig myenteric plexus. Neuroci Lett. 1986;68:57–62.
3. Morhenn VB, Waleh NS, Mansbridge JN, Unson D, Zolotorev A, Cline P, et al. Evidence for an NMDA receptor subunit in human keratinocytes and rat cardiocytes. Eur J Pharmacol. 1994;268:409–14.
4. Nahm WK, Philpot BD, Adams MM, Badiavas EV, Zhou LH, Butmarc J, et al. Significance of N-methyl-D-aspartate (NMDA) receptor-mediated signaling in human keratinocytes. J Cell Physiol. 2004;200:309–17.
5. Miglio G, Dianzani C, Fallarini S, Fantozzi R, Lombardi G. Stimulation of N-methyl-D-aspartate receptors modulates Jurkat T cell growth and adhesion to fibronectin. Biochem Biophys Res Commun. 2007;361:404–9.
6. Chen HJ, Fitzgerald R, Brown AT, Qureshi I, Breckenridge J, Kazi R, et al. Identification of a homocysteine receptor in the peripheral endothelium and its role in proliferation. J Vasc Surg. 2005;41:853–60.
7. McGee MA, Abdel-Rahman AA. Enhanced vascular neuronal nitric-oxide synthase-derived nitric-oxide production underlies the pressor response caused by peripheral N-methyl-D-aspartate receptor activation in conscious rats. J Pharmacol Exp Ther. 2012;342:461–71.
8. Leung JC, Travis BR, Verlander JW, Sandhu SK, Yang SG, Zea AH, et al. Expression and developmental regulation of the NMDA receptor subunits in the kidney and cardiovascular system. Am J Physiol Regul Integr Comp Physiol. 2002;283:R964–71.
9. Patton AJ, Genever PG, Birch MA, Suva LJ, Skerry TM. Expression of an N-methyl-D-aspartate-type receptor by human and rat osteoblasts and osteoclasts suggests a novel glutamate signaling pathway in bone. Bone. 1998;22:645–9.
10. Mentaverri R, Kamel S, Wattel A, Prouillet C, Sevenet N, Petit JP, et al. Regulation of bone resorption and osteoclast survival by nitric oxide: possible involvement of NMDA-receptor. J Cell Biochem. 2003;88:1145–56.
11. Seeber S, Becker K, Rau T, Eschenhagen T, Becker CM, Herkert M. Transient expression of NMDA receptor subunit NR2B in the developing rat heart. J Neurochem. 2000;75:2472–7.
12. Deng A, Valdivielso JM, Munger KA, Blantz RC, Thomson SC. Vasodilatory N-methyl-D-aspartate receptors are constitutively expressed in rat kidney. J Am Soc Nephrol. 2002;13:1381–4.

13. Gill SS, Pulido OM, Mueller RW, McGuire PF. Molecular and immunochemical characterization of the ionotropic glutamate receptors in the rat heart. Brain Res Bull. 1998;46:429–34.
14. Parisi E, Almadén Y, Ibarz M, Panizo S, Cardús A, Rodriguez M, et al. N-methyl-D-aspartate receptors are expressed in rat parathyroid gland and regulate PTH secretion. Am J Physiol Renal Physiol. 2009;296:F1291–6.
15. Gill SS, Pulido OM. Glutamate receptors in peripheral tissues: current knowledge, future research, and implications for toxicology. Toxicol Pathol. 2001;29:208–23.
16. Lin YJ, Bovetto S, Carver JM, Giordano T. Cloning of the cDNA for the human NMDA receptor NR2C subunit and its expression in the central nervous system and periphery. Brain Res Mol Brain Res. 1996;43:57–64.
17. Gonzalez-Cadavid NF, Ryndin I, Vernet D, Magee TR, Rajfer J. Presence of NMDA receptor subunits in the male lower urogenital tract. J Androl. 2000;21:566–78.
18. Ma MC, Huang HS, Chen YS, Lee SH. Mechanosensitive N-methyl-D-aspartate receptors contribute to sensory activation in the rat renal pelvis. Hypertension. 2008;52:938–U545.
19. Parisi E, Bozic M, Ibarz M, Panizo S, Valcheva P, Coll B, et al. Sustained activation of renal N-methyl-D-aspartate receptors decreases vitamin D synthesis: a possible role for glutamate on the onset of secondary HPT. Am J Physiol Endocrinol Metab. 2010;299:E825–31.
20. Bozic M, de Rooij J, Parisi E, Ruiz OM, Fernandez E, Valdivielso JM. Glutamatergic signaling maintains the epithelial phenotype of proximal tubular cells. J Am Soc Nephrol. 2011;22:1099–111.
21. Nasstrom J, Boo E, Stahlberg M, Berge OG. Tissue distribution of two NMDA receptor antagonists, [3H]CGS 19755 and [3H]MK-801, after intrathecal injection in mice. Pharmacol Biochem Behav. 1993;44:9–15.
22. Deng A, Thomson SC. Renal NMDA receptors independently stimulate proximal reabsorption and glomerular filtration. Am J Physiol Renal Physiol. 2009;296:F976–82.
23. Zhang C, Yi F, Xia M, Boini KM, Zhu Q, Laperle LA, et al. NMDA receptor-mediated activation of NADPH oxidase and glomerulosclerosis in hyperhomocysteinemic rats. Antioxid Redox Signal. 2010;13:975–86.
24. Anderson M, Suh JM, Kim EY, Dryer SE. Functional NMDA receptors with atypical properties are expressed in podocytes. Am J Physiol Cell Physiol. 2011;300:C22–32.
25. Giardino L, Armelloni S, Corbelli A, Mattinzoli D, Zennaro C, Guerrot D, et al. Podocyte glutamatergic signaling contributes to the function of the glomerular filtration barrier. J Am Soc Nephrol. 2009;20:1929–40.
26. Sproul AD, Steele SL, Thai TL, Yu SP, Klein JD, Sands JM, et al. N-methyl-D-aspartate receptor subunit NR3a expression and function in principal cells of the collecting duct. Am J Physiol Renal Physiol. 2011;301:F44–54.
27. Gu Y, Genever PG, Skerry TM, Publicover SJ. The NMDA type glutamate receptors expressed by primary rat osteoblasts have the same electrophysiological characteristics as neuronal receptors. Calcif Tissue Int. 2002;70:194–203.
28. Itzstein C, Cheynel H, Burt-Pichat B, Merle B, Espinosa L, Delmas PD, et al. Molecular identification of NMDA glutamate receptors expressed in bone cells. J Cell Biochem. 2001;82:134–44.
29. Ho ML, Tsai TN, Chang JK, Shao TS, Jeng YR, Hsu C. Down-regulation of N-methyl D-aspartate receptor in rat-modeled disuse osteopenia. Osteoporos Int. 2005;16:1780–8.
30. Chenu C, Serre CM, Raynal C, Burt-Pichat B, Delmas PD. Glutamate receptors are expressed by bone cells and are involved in bone resorption. Bone. 1998;22:295–9.
31. Kalariti N, Lembessis P, Koutsilieris M. Characterization of the glutamatergic system in MG-63 ostoblast-like osteosarcoma cells. Anticancer Res. 2004;24:3923–30.
32. Merle B, Itzstein C, Delmas PD, Chenu C. NMDA glutamate receptors are expressed by osteoclast precursors and involved in the regulation of osteoclastogenesis. J Cell Biochem. 2003;90:424–36.
33. Hinoi E, Fujimori S, Takarada T, Taniura H, Yoneda Y. Facilitation of glutamate release by ionotropic glutamate receptors in osteobasts. Biochem Biophys Res Commun. 2002;297:452–8.

34. Gu YC, Publicover SJ. Expression of functional metabotropic glutamate receptors primary cultured rat osteoblasts—cross-talk with N-methyl-D-aspartate receptors. J Biol Chem. 2000;275:34252–9.
35. Fujita H, Hinoi E, Nakatani E, Yamamoto T, Takarada T, Yoneda Y. Possible modulation of process extension by N-methyl-D-aspartate receptor expressed in osteocytic MLO-Y4 cells. J Pharmacol Sci. 2012;119:112–6.
36. Blantz RC, Deng A, Miracle CM, Thomson SC. Regulation of kidney function and metabolism: a question of supply and demand. Trans Am Clin Climatol Assoc. 2007;118:23–43.
37. Bądzyńska B, Zakrocka I, Sadowski J, Turski WA, Kompanowska-Jezierska E. Effects of systemic administration of kynurenic acid and glycine on renal haemodynamics and excretion in normotensive and spontaneously hypertensive rats. Eur J Pharmacol. 2014;743:37–41.
38. Slomowitz LA, Gabbai FB, Khang SJ, Satriano J, Thareau S, Deng AH, et al. Protein intake regulates the vasodilatory function of the kidney and NMDA receptor expression. Am J Physiol Regul Integr Comp Physiol. 2004;287:R1184–9.
39. Bellisle F, France B. Experimental studies of food choices and palatability responses in European subjects exposed to the Umami taste. Asia Pac J Clin Nutr. 2008;17(Suppl 1):376–9.
40. Mahieu S, Klug M, Millen N, Fabro A, Benmelej A, Contini MC. Monosodium glutamate intake affect the function of the kidney through NMDA receptor. Life Sci. 2016;149:114–9.
41. Kim EY, Anderson M, Dryer SE. Sustained activation of N-methyl-D-aspartate receptors in podoctyes leads to oxidative stress, mobilization of transient receptor potential canonical 6 channels, nuclear factor of activated T cells activation, and apoptotic cell death. Mol Pharmacol. 2012;82:728–37.
42. Kundu S, Pushpakumar SB, Tyagi A, Coley D, Sen U. Hydrogen sulfide deficiency and diabetic renal remodeling: role of matrix metalloproteinase-9. Am J Physiol Endocrinol Metab. 2013;304:E1365–78.
43. Kundu S, Pushpakumar S, Sen U. MMP-9- and NMDA receptor-mediated mechanism of diabetic renovascular remodeling and kidney dysfunction: hydrogen sulfide is a key modulator. Nitric Oxide. 2015;46:172–85.
44. Szaroma W, Dziubek K, Kapusta E. Effect of N-methyl-D-aspartic acid on activity of superoxide dismutase, catalase, glutathione peroxidase and reduced glutathione level in selected organs of the mouse. Acta Physiol Hung. 2014;101:377–87.
45. Frokiaer J. Collecting duct expression of N-methyl-D-aspartate receptor subtype NR3a regulates urinary concentrating capacity. Am J Physiol Renal Physiol. 2011;301:F42–3.
46. Franke WW, Schmid E, Osborn M, Weber K. Different intermediate-sized filaments distinguished by immunofluorescence microscopy. Proc Natl Acad Sci U S A. 1978;75:5034–8.
47. Leung JC, Marphis T, Craver RD, Silverstein DM. Altered NMDA receptor expression in renal toxicity: protection with a receptor antagonist. Kidney Int. 2004;66:167–76.
48. Leung JC, Ragland N, Marphis T, Silverstein DM. NMDA agonists and antagonists induce renal culture cell toxicity. Med Chem. 2008;4:565–71.
49. Basile AS, Huang JM, Xie C, Webster D, Berlin C, Skolnick P. N-methyl-D-aspartate antagonists limit aminoglycoside antibiotic-induced hearing loss. Nat Med. 1996;2:1338–43.
50. Puel JL, Ladrech S, Chabert R, Pujol R, Eybalin M. Electrophysiological evidence for the presence of NMDA receptors in the guinea-pig cochlea. Hear Res. 1991;51:255–64.
51. Segal JA, Harris BD, Kustova Y, Basile A, Skolnick P. Aminoglycoside neurotoxicity involves NMDA receptor activation. Brain Res. 1999;815:270–7.
52. Arora S, Kaur T, Kaur A, Singh AP. Glycine aggravates ischemia reperfusion-induced acute kidney injury through N-methyl-D-aspartate receptor activation in rats. Mol Cell Biochem. 2014;393:123–31.
53. Yang CC, Chien CT, Wu MH, Ma MC, Chen CF. NMDA receptor blocker ameliorates ischemia-reperfusion-induced renal dysfunction in rat kidneys. Am J Physiol Renal Physiol. 2008;294:F1433–40.
54. Pundir M, Arora S, Kaur T, Singh R, Singh AP. Effect of modulating the allosteric sites of N-methyl-D-aspartate receptors in ischemia-reperfusion induced acute kidney injury. J Surg Res. 2013;183:668–77.

55. Singh AP, Singh N, Bedi PM. Pioglitazone ameliorates renal ischemia reperfusion injury through NMDA receptor antagonism in rats. Mol Cell Biochem. 2016;417:111–8.
56. Lin CS, Hung SF, Huang HS, Ma MC. Blockade of the N-methyl-D-aspartate glutamate receptor ameliorates lipopolysaccharide-induced renal insufficiency. PLoS One. 2015;10:e0132204.
57. Cauli O, González-Usano A, Cabrera-Pastor A, Gimenez-Garzó C, López-Larrubia P, Ruiz-Sauri A, et al. Blocking NMDA receptors delays death in rats with acute liver failure by dual protective mechanisms in kidney and brain. Neuromolecular Med. 2014;16:360–75.
58. D'Amico M, Di Filippo C, Rossi F. Arrhythmias induced by myocardial ischaemia-reperfusion are sensitive to ionotropic excitatory amino acid receptor antagonists. Eur J Pharmacol. 1999;366:167–74.
59. Shi S, Liu T, Li Y, Qin M, Tang Y, Shen JY, et al. Chronic N-methyl-D-aspartate receptor activation induces cardiac electrical remodeling and increases susceptibility to ventricular arrhythmias. Pacing Clin Electrophysiol. 2014;37:1367–77. doi:10.1111/pace.12430.
60. Liu Y, Zhou L, Xu HF, Yan L, Ding F, Hao W, et al. A preliminary experimental study on the cardiac toxicity of glutamate and the role of alpha-amino-3-hydroxy-5-methyl-4-isoxazolepropionic acid receptor in rats. Chin Med J (Engl). 2013;126:1323–32.
61. Maldonado C, Soni CV, Todnem ND, Pushpakumar S, Rosenberger D, Givvimani S, et al. Hyperhomocysteinemia and sudden cardiac death: potential arrhythmogenic mechanisms. Curr Vasc Pharmacol. 2010;8:64–74.
62. Sun XL, Zhong JQ, Wang DG, Xu J, Su H, An CS, et al. Increasing glutamate promotes ischemia-reperfusion-induced ventricular arrhythmias in rats in vivo. Pharmacology. 2014;93:4–9.
63. Huang CF, Su MJ. Positive inotropic action of NMDA receptor antagonist (+)-MK801 in rat heart. J Biomed Sci. 1999;6:387–98.
64. Kleiger RE, Miller JP, Bigger JT, Moss AJ. Decreased heart rate variability and its association with increased mortality after acute myocardial infarction. Am J Cardiol. 1987;59:256–62.
65. Bennett AJ, DePetrillo PB. Differential effects of MK801 and lorazepam on heart rate variability in adolescent rhesus monkeys (Macoca mulatta). J Cardiovasc Pharmacol. 2005;45:383–8.
66. Aya AGM, Robert E, Bruelle P, Lefrant JY, Juan JM, Peray P, et al. Effects of ketamine on ventricular conduction, refractoriness, and wavelength - potential antiarrhythmic effects: a high-resolution epicardial mapping in rabbit hearts. Anesthesiology. 1997;87:1417–27.
67. Baczko I, Lepran I, Papp JG. Influence of anesthetics on the incidence of reperfusion-induced arrhythmias and sudden death in rats. J Cardiovasc Pharmacol. 1997;29:196–201.
68. Shi S, Liu T, Wang D, Zhang Y, Liang J, Yang B, et al. Activation of N-methyl-d-aspartate receptors reduces heart rate variability and facilitates atrial fibrillation in rats. Europace. 2016 May 11. pii: euw086.
69. Hageman GR, Simor T. Attenuation of the cardiac effects of cocaine by dizocilpine. Am J Physiol. 1993;264:H1890–5.
70. Gao X, Xu X, Pang J, Zhang C, Ding JM, Peng X, et al. NMDA receptor activation induces mitochondrial dysfunction, oxidative stress and apoptosis in cultured neonatal rat cardiomyocytes. Physiol Res. 2007;56:559–69.
71. Srejovic I, Jakovljevic V, Zivkovic V, Barudzic N, Radovanovic A, Stanojlovic O, et al. The effects of the modulation of NMDA receptors by homocysteine thiolactone and dizocilpine on cardiodynamics and oxidative stress in isolated rat heart. Mol Cell Biochem. 2015;401:97–105.
72. Tyagi N, Vacek JC, Givvimani S, Sen U, Tyagi SC. Cardiac specific deletion of N-methyl-D-aspartate receptor 1 ameliorates mtMMP-9 mediated autophagy/mitophagy in hyperhomocysteinemia. J Recept Signal Transduct. 2010;30:78–87.
73. Moshal KS, Tipparaju SM, Vacek TP, Kumar M, Singh M, Frank IE, et al. Mitochondrial matrix metalloproteinase activation decreases myocyte contractility in hyperhomocysteinemia. Am J Physiol Heart Circ Physiol. 2008;295:H890–7.
74. Tyagi N, Mishra PK, Tyagi SC. Homocysteine, hydrogen sulfide (H2S) and NMDA-receptor in heart failure. Indian J Biochem Biophys. 2009;46:441–6.

75. Meng L, Zhang ZX, Xu K, Qi GX. HIV-1 gp120 induces autophagy in cardiomyocytes via the NMDA receptor. Int J Cardiol. 2013;167:2517–23.
76. Meneghini A, Ferreira C, de Abreu LC, Valenti VE, Ferreira M, Filho CF, et al. Memantine prevents cardiomyocytes nuclear size reduction in the left ventricle of rats exposed to cold stress. Clinics. 2009;64:921–6.
77. Koizumi M, Komaba H, Fukagawa M. Parathyroid function in chronic kidney disease: role of FGF23-Klotho axis. Contrib Nephrol. 2013;180:110–23.
78. Ganor Y, Levite M. The neurotransmitter glutamate and human T cells: glutamate receptors and glutamate-induced direct and potent effects on normal human T cells, cancerous human leukemia and lymphoma T cells, and autoimmune human T cells. J Neural Transm. 2014;121:983–1006.
79. Divino Filho JC, Hazel SJ, Frst P, Bergstrm J, Hall K. Glutamate concentration in plasma, erythrocyte and muscle in relation to plasma levels of insulin-like growth factor (IGF)-I, IGF binding protein-1 and insulin in patients on haemodialysis. J Endocrinol. 1998;156:519–27.
80. Li JL, Cui B, Qi L, Li XY, Deng LF, Ning G, et al. NMDA enhances stretching-induced differentiation of osteobalsts through the ERK1/2 signaling pathway. Bone. 2008;43:469–75.
81. Mason DJ, Suva LJ, Genever PG, Patton AJ, Steuckle S, Hillam RA, et al. Mechanically regulated expression of a neural glutamate transporter in bone: a role for excitatory amino acids as osteotropic agents? Bone. 1997;20:199–205.
82. Hinoi E, Yoneda Y. Possible involvement of glutamatergic signaling machineries in pathophysiology of rheumatoid arthritis. J Pharmacol Sci. 2011;116:248–56.
83. Ralphe JC, Segar JL, Schutte BC, Scholz TD. Localization and function of the brain excitatory amino acid transporter type 1 in cardiac mitochondria. J Mol Cell Cardiol. 2004;37:33–41.
84. Hediger MA. Glutamate transporters in kidney and brain. Am J Physiol. 1999;277:F487–92.
85. Welbourne TC, Matthews JC. Glutamate transporters and renal function. Am J Physiol. 1999;277:F501–5.
86. Laketic-Ljubojevic I, Suva LJ, Maathuis FJ, Sanders D, Skerry TM. Functional characterization of N-methyl-D-aspartic acid-gated channels in bone cells. Bone. 1999;25:631–7.
87. Xue H, CJ F. New role of glutamate as an immunoregulator via glutamate receptors and transporters. Front Biosci (Schol Ed). 2011;1:1007–20.
88. Watanabe M, Mishina M, Inoue Y. Distinct gene expression of the N-methyl-D-aspartate receptor channel subunit in peripheral neurons of the mouse sensory ganglia and adrenal gland. Neurosci Lett. 1994;165:183–6.
89. Li JL, Zhao L, Cui B, Deng LF, Ning G, Liu JM. Multiple signaling pathways involved in stimulation of osteoblast differentiation by N-methyl-D-aspartate receptors activation in vitro. Acta Pharmacol Sin. 2011;32:895–903.

Chapter 4
Rapid Antidepressant Activity of Ketamine Beyond NMDA Receptor

Kenji Hashimoto

Abstract Multiple lines of evidence suggest that *N*-methyl-D-aspartate (NMDA) receptor plays a key role in the pathophysiology of depression and therapeutic mechanisms of antidepressants. The NMDA receptor antagonist ketamine is one of the most attractive antidepressants because it can produce rapid and sustained effects in patients with treatment-resistant depression. Recent meta-analyses have shown that the antidepressant effect of ketamine is more potent than that of other NMDA receptor antagonists [e.g., memantine, traxoprodil (CP-101,606), lanicemine (AZD6765), and rapastinel (GLYX-13)] in patients with depression. Ketamine is a racemic mixture containing equal parts of (*R*)-ketamine and (*S*)-ketamine (esketamine). In comparison with (*R*)-ketamine, esketamine shows approximately fourfold greater potency at NMDA receptor. We recently reported that in comparison with esketamine, (*R*)-ketamine shows greater potency and longer-lasting antidepressant effects in animal models of depression. Therefore, it is unlikely that NMDA receptor has a major role in the longer-lasting antidepressant effects of (*R*)-ketamine, although antagonism at this receptor may promote its rapid antidepressant activity. Unlike esketamine, (*R*)-ketamine does not induce psychotomimetic side effects or have abuse potential in rodents. In this chapter, we discuss the role of NMDA receptor in the antidepressant activities of ketamine and its enantiomers.

Keywords Antidepressant • Brain-derived neurotrophic factor • Ketamine • Esketamine • (*R*)-ketamine

Abbreviations

AMPA	α-Amino-3-hydroxy-5-methyl-4-isoxazolepropionic acid
BDNF	Brain-derived neurotrophic factor

K. Hashimoto (✉)
Division of Clinical Neuroscience, Chiba University Center for Forensic Mental Health, 1-8-1 Inohana, Chiba 260-8670, Japan
e-mail: hashimoto@faculty.chiba-u.jp

K. Hashimoto (ed.), *The NMDA Receptors*, The Receptors 30,
DOI 10.1007/978-3-319-49795-2_4

BPRS	Brief Psychiatric Rating Scale
CADSS	Clinician-Administered Dissociative States Scale
CNS	Central nervous system
FDA	Food Drug Administration
GLT-1	Glutamate transporter 1
mTORC1	Mammalian target of rapamycin complex 1
NMDA	*N*-methyl-D-aspartate
PCP	Phencyclidine
PV	Parvalbumin
TrkB	Tropomyosin receptor kinase B

4.1 Introduction

Glutamate (L-glutamic acid) is one of the major excitatory neurotransmitters in the mammalian central nervous system (CNS). Multiple lines of evidence suggest that glutamatergic neurotransmission via *N*-methyl-D-aspartate (NMDA) receptor plays a key role in the pathophysiology of depression and mechanisms of action of antidepressants [1–8].

In 2000, Burman et al. (Yale University) reported a rapid antidepressant effect of the NMDA receptor antagonist ketamine in patients with depression [9]. Subsequently, randomized, placebo-controlled studies demonstrated that ketamine produced rapid and sustained antidepressant effects in treatment-resistant patients with major depression and bipolar depression [10, 11]. Furthermore, ketamine showed antidepressant effects in electroconvulsive therapy (ECT)-resistant patients with depression [12]. These reports and many subsequent clinical studies make ketamine an attractive rapid-onset therapeutic drug for treatment-resistant depression, although its clinical application may be limited owing to its propensity of causing psychotomimetic effects [13–21].

In this chapter, I discuss the role of NMDA receptor in the antidepressant activities of ketamine.

4.2 History of Ketamine and Its Abuse Liability

Ketamine (formerly CI-581) was first synthesized in 1962 by Calvin L. Stevens (Wayne State University). After preclinical research, an intravenous subanesthetic dose of ketamine was introduced for testing in human prisoners in 1964. Edward F. Domino (Michigan University) and his wife Toni asserted that ketamine was "dissociative anesthesia" [22]. They demonstrated ketamine's short duration of action and low behavioral toxicity, which made it a more favorable choice compared with phencyclidine (PCP; formerly CI-395) as a dissociative anesthetic drug. Since its

Food and Drug Administration (FDA) approval in 1970, ketamine has been widely used as a dissociative anesthetic [22, 23].

During the late 1960s and early 1970s, many drugs were used by young people as part of "make love, not war" protests against the U.S. war in Vietnam. Ketamine is widely used in veterinary medicine, but sterile ketamine vials intended for veterinary use were diverted for recreational use [22]. Because of its lower potency and shorter duration of action, ketamine ("special K") is associated with less severe psychiatric problems than PCP ("angel dust") [24]. In some countries, ketamine is the most commonly abused drug, and the prevalence of health and social problems is associated with ketamine abuse [25]. Thus, ketamine is a scheduled drug, the use of which should be restricted because of its abuse liability.

4.3 Mechanism of Action of Ketamine

Lodge's group (Royal Veterinary College) reported for the first time that ketamine and PCP are selective NMDA receptor antagonists [24, 26]. The schizophrenia-like actions of PCP detected in Luby's study are well known [27]. Subsequently, ketamine caused schizophrenia-like symptoms in humans [22]. Thus, the psychosis caused by PCP and ketamine was the most closely related to schizophrenia [24, 27, 28]. In 1994, Krystal et al. (Yale University) reported that ketamine has positive and negative effects, including cognitive impairment, in healthy control subjects [29]. Taken together, these findings suggest the NMDA receptor hypofunction hypothesis in schizophrenia [30–34].

Ketamine is a racemic mixture of (*S*)-ketamine (esketamine) and (*R*)-ketamine (Fig. 4.1.). Esketamine ($K_i = 0.30\ \mu M$ for NMDA receptor) has approximately four-fold higher affinity for NMDA receptor relative to (*R*)-ketamine ($K_i = 1.4\ \mu M$) (Fig. 4.1.) [35]. NMDA receptors are tetrameric combinations usually comprising two GluN1 and two GluN2 subunits, with four possible genes (A–D) encoding the latter.

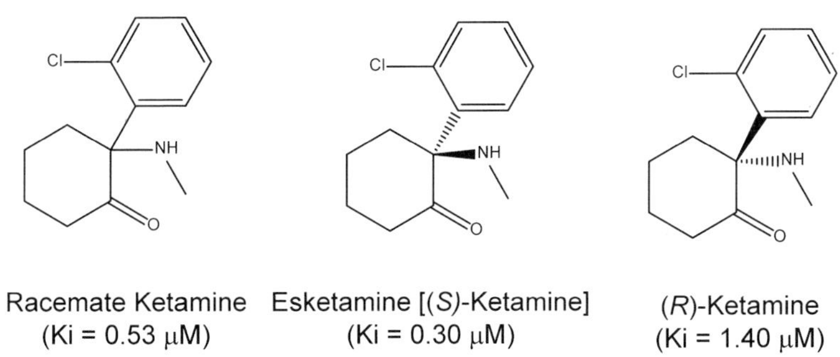

Fig. 4.1 Chemical structure of ketamine, ketamine enantiomers. The value in the parenthesis is the Ki value for NMDA receptor [35]

Esketamine was found to be approximately nine times less potent on GluN1/GluN2A than on GluN1/GluN2B-D [36].

Ketamine has several mechanisms of action besides NMDA receptor antagonism. Esketamine is two to three times more potent than (*R*)-ketamine at μ, κ, and δ opioid receptors [37, 38], although the opioid receptor antagonist naloxone did not block ketamine's action in humans [39]. Furthermore, ketamine has moderate affinity for the dopamine D_2 receptor [40, 41], although the affinity of the two enantiomers at D_2 receptors has not been investigated. (*R*)-ketamine has weak affinity for the sigma-1 receptor, at which only negligible binding of esketamine occurs [42]. However, the precise mechanisms of ketamine's activities are currently unknown.

4.4 Pharmacokinetic Profile of Ketamine

Because of its extensive first-pass metabolism, the oral bioavailability of ketamine is poor. Thus, it has been used via intravenous, intramuscular, and topical routes. Sublingual and nasal formulations of ketamine have also been developed [43]. Ketamine has short blood α and β $t_{1/2}$ of approximately 7 min and 2–4 h, respectively. Its metabolites (norketamine and dehydronorketamine) appear in venous blood approximately 10 and 30 min after administration. Ketamine is metabolized in the liver by CYP3A4 (major), CYP2B6 (minor), and CYP2C9 (minor) isoenzymes into norketamine (through N-demethylation) and finally dehydronorketamine (Fig. 4.2.). Esketamine and (*R*)-ketamine have similar pharmacokinetic profiles [22]. In addition, (*R*)-ketamine is not formed after the intravenous administration of esketamine in humans, indicating the lack of their interconversion [44, 45].

In human volunteers, intravenous esketamine (0.15 mg/kg) is more potent than (*R*)-ketamine (0.5 mg/kg) as an analgesic [46]. However, esketamine produced 1.6 times greater altered body image and changes in hearing, 2.5 times greater feelings of unreality, and 4 times more reduced visual acuity. Thus, it is likely that esketamine because of having greater potency at NMDA receptors has more unwanted psychiatric side effects than (*R*)-ketamine [22], supporting the hypothesis of NMDA receptor hypofunction in psychosis [30–34].

4.5 Antidepressant Effects of Ketamine Racemate

As mentioned in the introduction, Robert Burman et al. (Yale University) reported a rapid antidepressant effect in patients with depression [9]. Subsequently, randomized, placebo-controlled studies demonstrated that ketamine produced rapid antidepressant effects in patients with treatment-resistant and bipolar depression [10, 11]. A randomized, active placebo (midazolam) control study showed that ketamine had greater improvement in the depression score than the midazolam group 24 h after infusion [47]. Singh et al. [48] reported that two- and three-time weekly infusions

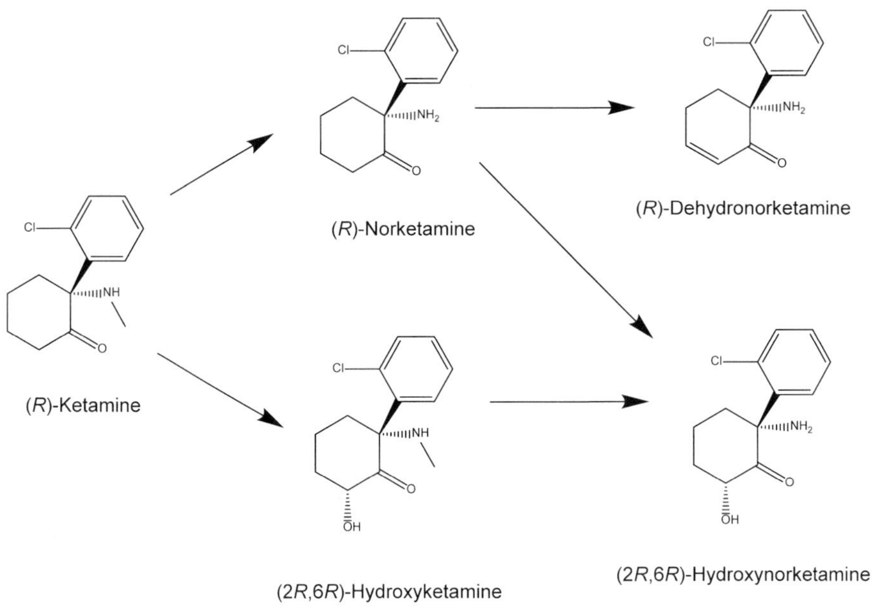

Fig. 4.2 Metabolism of (*R*)-ketamine to (*R*)-norketamine, (*R*)-dehydronorketamine, (2*R*,6*R*)-hydroxyketamine and (2*R*,6*R*)-hydroxynorketamine

of ketamine (0.5 mg/kg) maintained antidepressant efficacy over 15 days, although dissociative symptoms occurred transiently and were attenuated by repeated dosing. Recent meta-analyses have demonstrated that non-ketamine NMDA receptor antagonists [e.g., memantine, traxoprodil (CP-101,606), lanicemine (AZD6765), and rapastinel (GLYX-13)] have smaller effects than ketamine racemate, although the reason underlying this remains unclear [49, 50].

4.6 Mechanisms of Ketamine's Antidepressant Activity

Although ketamine's rapid-onset antidepressant activity is well known, the cellular and molecular mechanisms underlying this remain unclear [51–55]. The rapid antidepressant action of ketamine is currently thought to occur via the blockade of NMDA receptors located on inhibitory γ-aminobutylic acid (GABA)ergic neurons. This causes disinhibition of the pyramidal cells, resulting in a burst of glutamate transmission. Increased glutamate release activates α-amino-3-hydroxy-5-methyl-4-isoxazolepropionic acid (AMPA) receptors, resulting in depolarization and calcium influx. Depolarization of the cell induces the release of brain-derived neurotrophic factor (BDNF) and activation of the mammalian target of rapamycin complex 1 (mTORC1) signaling pathway. Finally, the stimulation of mTORC1 increases the synthesis of synaptic proteins, which results in increased number and function of spine synapses [56–58]. A recent study showed that the antidepressant effect of ketamine via mTOR signaling is mediated by the inhibition of nitrergic Rheb degradation [59].

BDNF and its receptor, tropomyosin receptor kinase B (TrkB), play a role in the pathophysiology of depression and in the therapeutic mechanisms of antidepressants [60–63]. In preclinical models, we reported that regional differences in BDNF, its precursor proBDNF, and BDNF pro-peptide confer resilience to stress [64, 65]. Furthermore, TrkB agonist and antagonist showed a rapid antidepressant effect in inflammation [66, 67], social defeat stress [68], and learned helplessness [69] models of depression. It has been reported that BDNF plays a role in the antidepressant effect of ketamine because its antidepressant effects could be antagonized by TrkB antagonist [70, 71]. Taking these findings together, it is likely that BDNF–TrkB signaling plays a key role in the antidepressant effects of ketamine.

The protein p11 (also known as S100A10), a member of the S100 EF-hand protein family, is widely expressed in several brain regions that are implicated in the pathophysiology of depression, including the hippocampus and frontal cortex. Accumulating evidence suggests a key role of p11 in the pathophysiology of depression [72]. For example, a recent study showed that hippocampal p11 played a key role in the sustained antidepressant effect of ketamine in a chronic unpredictable mild stress model of depression [73].

Growing evidence suggests that downregulated clearance of glutamate and signaling pathways involving BDNF–TrkB signaling play a role in the morphological changes within the hippocampus of patients with depression. A recent study showed that the regulation of glutamate transporter 1 (GLT-1) on astrocytes, responsible for 90 % of glutamate reuptake from the synapse, through BDNF–TrkB signaling is involved in ketamine's antidepressant activity [74].

Panos Zanos et al. (University of Maryland) recently reported that the metabolism of ketamine to (2*S*,6*S*; 2*R*,6*R*)-hydroxynorketamine (Fig. 4.2.) is essential for its antidepressant effects [75]. These antidepressant activities are independent of NMDA receptor inhibition but involve early and sustained activation of AMPA receptors. This study suggests a novel mechanism underlying the antidepressant properties of ketamine; this mechanism is important for the development of next-generation, rapid-acting antidepressants. In contrast, we recently reported that a bilateral infusion of (*R*)-ketamine into the medial prefrontal cortex caused antidepressant effects in the rat learned helplessness model [76], indicating that a direct antidepressant action of (*R*)-ketamine itself. Very recently, we reported that (*R*)-ketamine showed greater potency and longer lasting antidepressant effects than its metabolite (2*R*,6*R*)-hydroxynorketamine in inflammation and social defeat stress models [77].

4.7 Antidepressant Effects of Esketamine

Many scientists believe that NMDA receptor antagonism plays a key role in the mechanisms of ketamine's antidepressant activity. Considering the high affinity of esketamine at NMDA receptor [approximately four times more potent than

(*R*)-ketamine], the company Johnson & Johnson has been developing a method for the intranasal administration of esketamine as a treatment for depression. Intranasal esketamine received the breakthrough treatment designation from the US FDA. In addition, Singh et al. [78] reported a rapid-onset antidepressant effect of intravenous esketamine infusion in treatment-resistant patients with depression, although the Brief Psychiatric Rating Scale (BPRS) score and the Clinician-administered Dissociative States Scale (CADSS) score peaked 40 min after esketamine infusion (0.20 or 0.40 mg/kg for 40 min). It is also likely that the potency of the antidepressant effect of intranasal ketamine administration is lower than that of the effect of intravenous ketamine infusion [79].

4.8 Antidepressant Effects of *R*-Ketamine

Because non-ketamine NMDA receptor antagonists have smaller effects than ketamine [49, 50], we hypothesized that NMDA receptor may not play a key role in the antidepressant effects of ketamine. Given the different affinities of the two ketamine enantiomers for NMDA receptor (Fig. 4.1.) [22, 35], we compared the antidepressant effects and side effect profiles of these two enantiomers in rodents.

We found that (*R*)-ketamine showed greater potency and longer-lasting antidepressant effects than esketamine in animal models of depression, including neonatal dexamethasone exposure, repeated social defeat stress, and learned helplessness [80, 81]. Therefore, it is unlikely that NMDA receptor has a major role in the long-lasting antidepressant effects of (*R*)-ketamine, although antagonism at this receptor may promote its rapid antidepressant activity [81]. Our findings have been replicated by a recent study [75]. Unlike esketamine, (*R*)-ketamine does not induce psychotomimetic side effects or abuse potential in rodents [81, 82]. Furthermore, we reported that a single dose of esketamine (10 mg/kg) but not (*R*)-ketamine (10 mg/kg) resulted in the loss of parvalbumin (PV) immunoreactivity in mouse brain regions, such as the prefrontal cortex [81], suggesting that the loss of PV-positive cells is associated with ketamine-induced psychotomimetic effects.

A recent study using [^{11}C]raclopride and positron emission tomography showed a marked reduction of dopamine $D_{2/3}$ receptor binding in the monkey striatum after a single infusion of esketamine (0.5 mg/kg, 40 min) but not (*R*)-ketamine (0.5 mg/kg, 40 min) [83]. Singh et al. [78] reported a rapid-onset antidepressant effect of esketamine in patients with treatment-resistant depression, although the BPRS and CADSS scores peaked 40 min after esketamine infusion (0.20 or 0.40 mg/kg for 40 min). Considering the role of dopamine release in psychosis, it is likely that the marked release of dopamine from presynaptic terminals in the striatum is associated with the psychotomimetic side effects in humans after an infusion of ketamine or esketamine. It is well known that psychosis induced by NMDA receptor antagonists such as ketamine and PCP could be associated with NMDA receptor antagonism [24, 30], suggesting that the psychotomimetic effects of ketamine and esketamine are associated with their antagonism at NMDA receptor.

Studies using repeated ketamine (or esketamine) infusions resulted in significant antidepressant effects with an extended median time to recurrence of depressive symptoms in a 4-week open-label study [13], an 18-day open-label study [84], a 12-month, naturalistic, three-patient case series [85], a four-open-label-injection study [78], and a double-blind placebo-controlled study [48]. However, psychotomimetic side effects were shown after each infusion of ketamine or esketamine [13, 48, 84]. There were no differences in dissociative, psychotomimetic, or high feelings between responders and non-responders [84], suggesting that ketamine's antidepressant effects are not associated with psychotomimetic effects. We recently reported that repeated, intermittent administration of esketamine (10 mg/kg, once per week for 8 weeks) but not (*R*)-ketamine led to loss of PV immunoreactivity in the prefrontal cortex of mouse brain [86]. Because such loss of PV immunoreactivity in the prefrontal cortex may be associated with psychosis and γ-oscillation deficits in schizophrenia [87, 88], repeated administration of esketamine or ketamine may have long-lasting detrimental side effects in the prefrontal cortex of humans. Thus, it seems that the loss of PV immunoreactivity in the prefrontal cortex is associated with NMDA receptor antagonism. Taking these findings together, it is likely that the repeated intermittent use of (*R*)-ketamine is safer than that of esketamine or ketamine in the treatment of depression [14, 17, 18].

Rapastinel (formerly GLYX-13), a partial agonist at glycine site of the NMDA receptor, shows antidepressant-like effects without ketamine-like side effects in animal models [89]. A recent double-blind, placebo-controlled study demonstrated that a single intravenous (i.v.) infusion of rapastinel (5 or 10 mg/kg) produced rapid and sustained antidepressant effects in depressed patients who had not responded to another antidepressant, and that this drug did not elicit psychotomimetic or other significant side effects [90]. The Phase III study of rapastinel (Allergan) received the Breakthrough Therapy designation from the U.S. FDA for adjunctive treatment of major depression. Very recently, we reported that (*R*)-ketamine is a longer-lasting antidepressant compared with rapastinel in social defeat stress model of depression [91].

These findings suggest that in comparison with esketamine and rapastinel, (*R*)-ketamine shows greater potency and longer-lasting antidepressant effects in animal models of depression. (*R*)-ketamine has fewer psychotomimetic side effects and lower abuse potential than esketamine. However, further detailed studies on the precise molecular and cellular mechanisms of (*R*)-ketamine's antidepressant effect are needed.

4.9 Conclusions and Perspectives

There is an urgent need for rapid-onset antidepressants for treatment-resistant depression. A number of clinical studies have demonstrated that ketamine has rapid-onset and sustained antidepressant effects in patients with treatment-resistant depression, although psychotomimetic effects of ketamine infusion have also been identified. Although ketamine has not yet been approved for use in depression, it has been widely used as an off-label approach in US. Because preclinical data suggests that repeated infusions of ketamine leads to detrimental side effects on the brain, careful screening, management, and follow-up of depressed patients who have received repeated ketamine therapy will be necessary.

In conclusion, the use of (*R*)-ketamine for treatment-resistant depression should provide a new therapeutic approach by reducing the detrimental side effects of racemate ketamine [14, 17, 18, 92] since possible advantages and disadvantages in the potential clinical use of racemate ketamine are pointed [93, 94].

Acknowledgements This study is supported by the Strategic Research Program for Brain Sciences from Japan Agency for Medical Research and Development, AMED.

Conflict of Interest Dr. Hashimoto is an inventor on a filed patent application on "The use of (*R*)-ketamine in the treatment of psychiatric diseases" by Chiba University. Dr. Hashimoto has received research support from Dainippon Sumitomo, Mochida, Otsuka, and Taisho.

References

1. Hashimoto K, Sawa A, Iyo M. Increased levels of glutamate in brains from patients with mood disorders. Biol Psychiatry. 2007;62(11):1310–6.
2. Sanacora G, Zarate CA, Krystal JH, Manji HK. Targeting the glutamatergic system to develop novel, improved therapeutics for mood disorders. Nat Rev Drug Discov. 2008;7(5):426–37.
3. Hashimoto K. Emerging role of glutamate in the pathophysiology of major depressive disorder. Brain Res Rev. 2009;61(2):105–23.
4. Hashimoto K. The role of glutamate on the action of antidepressants. Prog Neuropsychopharmacol Biol Psychiatry. 2011;35(7):1558–68.
5. Tokita K, Yamaji T, Hashimoto K. Roles of glutamate signaling in preclinical and/or mechanistic models of depression. Pharmacol Biochem Behav. 2012;100(4):688–704.
6. Hashimoto K, Malchow B, Falkai P, Schmitt A. Glutamate modulators as potential therapeutic drugs in schizophrenia and affective disorders. Eur Arch Psychiatry Clin Neurosci. 2013;263(5):367–77.
7. Dang YH, Ma XC, Zhang JC, Ren Q, Wu J, Gao CG, Hashimoto K. Targeting of NMDA receptors in the treatment of major depression. Curr Pharm Des. 2014;20(32):5151–9.
8. Ohgi Y, Futamura T, Hashimoto K. Glutamate signaling in synaptogenesis and NMDA receptors as potential therapeutic targets for psychiatric disorders. Curr Mol Med. 2015;15(3):206–21.
9. Berman RM, Cappiello A, Anand A, Oren DA, Heninger GR, Charney DS, Krystal JH. Antidepressant effects of ketamine in depressed patients. Biol Psychiatry. 2000;47(4):351–4.
10. Zarate Jr CA, Singh JB, Carlson PJ, Brutsche NE, Ameli R, Luckenbaugh DA, Charney DS, Manji HK. A randomized trial of an *N*-methyl-D-aspartate antagonist in treatment-resistant major depression. Arch Gen Psychiatry. 2006;63(8):856–64.
11. Diazgranados N, Ibrahim L, Brutsche NE, Newberg A, Kronstein P, Khalife S, Kammerer WA, Quezado Z, Luckenbaugh DA, Salvadore G, Machado-Vieira R, Manji HK, Zarate Jr CA. A randomized add-on trial of an *N*-methyl-D-aspartate antagonist in treatment-resistant bipolar depression. Arch Gen Psychiatry. 2010;67(8):793–802.
12. Ibrahim L, Diazgranados N, Luckenbaugh DA, Machado-Vieira R, Baumann J, Mallinger AG, Zarate Jr CA. Rapid decrease in depressive symptoms with an N-methyl-D-aspartate antagonist in

ECT-resistant major depression. Prog Neuropsychopharmacol Biol Psychiatry. 2011;35(4):1155–9.
13. Aan het Rot M, Collins KA, Murrough JW, Perez AM, Reich DL, Charney DS, Mathew SJ. Safety and efficacy of repeated-dose intravenous ketamine for treatment-resistant depression. Biol Psychiatry. 2010;67(2):139–45.
14. Hashimoto K. The *R*-stereoisomer of ketamine as an alternative for ketamine for treatment-resistant major depression. Clin Psychopharmacol Neurosci. 2014;12(1):72–3.
15. Zhang MW, Ho RC. Ketamine's potential as a rapid antidepressant was overplayed. BMJ. 2015;351:h4467.
16. Zhang MW, Ho R. Critical appraisal of existing ketamine trials: existing limitations and limited applicability for treatment. Am J Psychiatry. 2016;173(4):431.
17. Hashimoto K. *R*-ketamine: a rapid-onset and sustained antidepressant without risk of brain toxicity. Psychol Med. 2016;46(11):2449–51.
18. Hashimoto K. Detrimental side effects of repeated ketamine infusions in the brain. Am J Psychiatry. 2016;173(10):1044–5.
19. Abdallah CG, Adams TG, Kelmendi B, Esterlis I, Sanacora G, Krystal JH. Ketamine's mechanism of action: a path to rapid-acting antidepressants. Depress Anxiety. 2016;33(8):689–97.
20. Schwartz J, Murrough JW, Iosifescu DV. Ketamine for treatment-resistant depression: recent developments and clinical applications. Evid Based Ment Health. 2016;19(2):35–8.
21. Bobo WV, Voort JL, Croarkin PE, Leung JG, Tye SJ, Frye MA. Ketamine for treatment-resistant unipolar and bipolar depression: critical review and implications for clinical practice. Depress Anxiety. 2016;33(8):698–710.
22. Domino EF. Taming the ketamine tiger. 1965. Anesthesiology. 2010;113:678–84.
23. White PF, Way WL, Trevor AJ. Ketamine – its pharmacology and therapeutic use. Anesthesiology. 1982;56:119–36.
24. Lodge D, Mercier MS. Ketamine and phencyclidine: the good, the bad and the unexpected. Bri J Pharmacol. 2015;172(17):4254–76.
25. Zhang MW, Harris KM, Ho RC. Is off-label repeat prescription of ketamine as a rapid antidepressant safe? Controversies, ethical concerns, and legal implications. BMC Med Ethics. 2016;17:4.
26. Anis NA, Berry SC, Burton NR, Lodge D. The dissociative anaesthetics, ketamine and phencyclidine, selectively reduce excitation of central mammalian neurones by N-methyl-aspartate. Br J Pharmacol. 1983;79(2):565–75.
27. Luby ED, Cohen BD, Rosenbaum G, Gottlieb JS, Kelley R. Study of a new schizophrenomimetic drug – senyl. AMA Arch Neurol Psychiatry. 1959;81:363–9.
28. Domino EF, Chodoff P, Corssen G. Pharmacologic effects of CI-581, a new dissociative anesthetic, in man. Clin Pharmacol Ther. 1965;6:279–91.
29. Krystal JH, Karper LP, Seibyl JP, Freeman GK, Delaney R, Bremner JD, Heninger GR, Bowers Jr MB, Charney DS. Subanesthetic effects of the noncompetitive NMDA antagonist, ketamine, in humans. Psychotomimetic, perceptual, cognitive, and neuroendocrine responses. Arch Gen Psychiatry. 1994;51(3):199–214.
30. Javitt DC, Zukin SR. Recent advances in the phencyclidine model of schizophrenia. Am J Psychiatry. 1991;148(10):1301–8.
31. Olney JW, Farber NB. Glutamate receptor dysfunction and schizophrenia. Arch Gen Psychiatry. 1995;52(12):998–1007.
32. Gonzalez-Burgos G, Lewis DA. NMDA receptor hypofunction, parvalbumin-positive neurons and cortical gamma osillations in schizophrenia. Schizophr Bull. 2012;38(5):950–7.
33. Javitt DC, Zukin SR, Heresco-Levy U, Umbricht D. Has an angel shown the way? Etiological and therapeutic implications of the PCP/NMDA model of schizophrenia. Schizophr Bull. 2012;38(5):958–66.
34. Hashimoto K. Targeting of NMDA receptors in new treatments for schizophrenia. Expert Opin Ther Targets. 2014;18(9):1049–63.
35. Ebert B, Mikkelsen S, Thorkildsen C, Borgbjerg FM. Norketamine, the main metabolites of ketamine, is a non-competitive NMDA receptor antagonist in the rat cortex and spinal cord. Eur J Pharmacol. 1997;333(1):99–104.

36. Dravid SM, Erreger K, Yuan H, Nicholson K, Le P, Lyuboslavsky P, Almonte A, Murray E, Mosely C, Barber J, French A, Balster R, Murray TF, Traynelis SF. Subunit-specific mechanisms and proton sensitivity of NMDA receptor channel block. J Physiol. 2007;581(Pt 1):107–28.
37. Hirota K, Okawa H, Appadu BL, Grandy DK, Devi LA, Lambert DG. Stereoselective interaction of ketamine with recombinant mu, kappa, and delta opioid receptors expressed in Chinese hamster ovary cells. Anesthesiology. 1999;90(1):174–82.
38. Hirota K, Lambert DG. Ketamine: new uses for an old drug? Bri J Anaesth. 2011;107(2): 123–6.
39. Mikkelsen S, Ilkjaer S, Brennum J, Borgbjerg FM, Dahl JB. The effect of naloxone on ketamine-induced effects on hyperalgesia and ketamine-induced side effects in humans. Anesthesiology. 1999;90(6):1539–45.
40. Kapur S, Seeman P. NMDA receptor antagonists ketamine and PCP have direct effects on the dopamine D_2 and serotonin 5-HT_2 receptors – implications for models of schizophrenia. Mol Psychiatry. 2002;7(8):837–44.
41. Seeman P, Ko F, Tallerico T. Dopamine receptor contribution to the action of PCP, LSD, and ketamine psychotomimetics. Mol Psychiatry. 2005;10(9):837–83.
42. Øye I, Hustveit O, Maurset A, Ratti Moberg E, Paulsen O, Skoglund LA. The chiral forms of ketamine as probes for NMDA-receptor function in humans. In: Kameyama T, Nabeshima T, Domino EF, editors. NMDA receptor related agents: biochemistry, pharmacology and behavior. Ann Arbor: NPP Books; 1991. p. 381–9.
43. Peltoniemi MA, Hagelberg NM, Olkkola KT, Saari TI. Ketamine: a review of clinical pharmacokinetics and pharmacodynamics in anesthesia and pain therapy. Clin Pharmacokinet. 2016;55(9):1059–77.
44. Geisslinger G, Hering W, Thomann P, Knoll R, Kamp HD, Brune K. Pharmacokinetics and pharmacodynamics of ketamine enantiomers in surgical patients using a stereoselective analytical method. Br J Anaesth. 1993;70(6):666–71.
45. Ihmsen H, Geisslinger G, Schüttler J. Stereoselective pharmacokinetics of ketamine: R(-)-ketamine inhibits the elimination of S(+)-ketamine. Clin Pharmacol Ther. 2001;70(5):431–8.
46. Oye I, Paulsen O, Maurset A. Effects of ketamine on sensory perception: evidence for a role of N-methyl-D-aspartate receptors. J Pharmacol Exp Ther. 1992;260(3):1209–13.
47. Murrough JW, Iosifescu DV, Chang LC, Al Jurdi RK, Green CE, Perez AM, Iqbal S, Pillemer S, Foulkes A, Shah A, Charney DS, Mathew SJ. Antidepressant efficacy of ketamine in treatment-resistant major depression: a two-site randomized controlled trial. Am J Psychiatry. 2013;170(10):1134–42.
48. Singh JB, Fedgchin M, Daly EJ, De Boer P, Cooper K, Lim P, Pinter C, Murrough JW, Sanacora G, Shelton RC, Kurian B, Winokur A, Fava M, Manji H, Drevets WC, Van Nueten L. A double-blind, randomized, placebo-controlled, dose-frequency study of intravenous ketamine in patients with treatment-resistant depression. Am J Psychiatry. 2016;173(8):816–26.
49. Newport DJ, Carpenter LL, McDonald WM, Potash JB, Tohen M. Nemeroff CB; APA Council of Research Task Force on Novel Biomarkers and Treatments. Ketamine and other NMDA antagonists: early clinical trials and possible mechanisms in depression. Am J Psychiatry. 2015;172(10):950–66.
50. Kishimoto T, Chawla JM, Hagi K, Zarate CA, Kane JM, Bauer M, Correll CU. Single-dose infusion ketamine and non-ketamine *N*-methyl-D-aspartate receptor antagonists for unipolar and bipolar depression: a meta-analysis of efficacy, safety and time trajectories. Psychol Med. 2016;46(7):1459–72.
51. Arnold C. The promise and perils of ketamine research. Lancet. 2013;12(10):940–1.
52. Rush AJ. Ketamine for treatment-resistant depression: ready or not for clinical use? Am J Psychiatry. 2013;170(10):1079–81.
53. Schwatzberg AF. A word to the wide about ketamine. Am J Psychiatry. 2014;171(3):262–4.
54. Kavalali ET, Monteggia LM. Synaptic mechanisms underlying rapid antidepressant action of ketamine. Am J Psychiatry. 2012;169(11):1150–6.

55. Lester HA, Lavis LD, Dougherty DA. Ketamine inside neurons? Am J Psychiatry. 2015;172(11):1064–6.
56. Duman RS, Aghajanian GK. Synaptic dysfunction in depression: potential therapeutic targets. Science. 2012;338(6103):68–72.
57. Dwyer JM, Duman RS. Activation of mammalian target of rapamycin and synaptogenesis: role in the actions of rapid-acting antidepressants. Biol Psychiatry. 2013;73(12):1189–98.
58. Gerhard DM, Wohleb ES, Duman RS. Emerging treatment mechanisms for depression: focus on glutamate and synaptic plasticity. Drug Discov Today. 2016;21(3):454–64.
59. Harraz MM, Tyagi R, Cortés P, Snyder SH. Antidepressant action of ketamine via mTOR is mediated by inhibition of nitrergic Rheb degradation. Mol Psychiatry. 2016;21(3):313–9.
60. Hashimoto K, Shimizu E, Iyo M. Critical role of brain-derived neurotrophic factor in mood disorders. Brain Res Brain Res Rev. 2004;45(2):104–14.
61. Hashimoto K. Brain-derived neurotrophic factor as a biomarker for mood disorders: an historical overview and future directions. Psychiatry Clin Neurosci. 2010;64(4):341–57.
62. Hashimoto K. Sigma-1 receptor chaperone and brain-derived neurotrophic factor: emerging links between cardiovascular disease and depression. Prog Neurobiol. 2013;100:15–9.
63. Hashimoto K. Brain-derived neurotrophic factor (BDNF) and its precursor proBDNF as diagnostic biomarkers for major depressive disorder and bipolar disorder. Eur Arch Psychiatry Clin Neurosci. 2015;265(1):83–4.
64. Yang C, Shirayama Y, Zhang JC, Ren Q, Hashimoto K. Regional differences in brain-derived neurotrophic factor levels and dendritic spine density confer resilience to inescapable stress. Int J Neuropsychopharmacol. 2015;18(7):pyu121.
65. Yang B, Yang C, Ren Q, Zhang JC, Chen QX, Shirayama Y, Hashimoto K. Regional differences in the expression of brain-derived neurotrophic factor (BDNF) pro-peptide, proBDNF and preproBDNF in the brain confer stress resilience. Eur Arch Psychiatry Clin Neurosci. 2016;266(8):765–9.
66. Zhang JC, Wu J, Fujita Y, Yao W, Ren Q, Yang C, Li SX, Shirayama Y, Hashimoto K. Antidepressant effects of TrkB ligands on depression-like behavior and dendritic changes in mice after inflammation. Int J Neuropsychopharmacol. 2015;18(4):pyu077.
67. Zhang JC, Yao W, Hashimoto K. Brain-derived neurotrophic factor (BDNF) - TrkB signaling in inflammation-related depression and potential therapeutic targets. Curr Neuropharmacol. 2016;14(7):721–31.
68. Zhang JC, Yao W, Dong C, Yang C, Ren Q, Ma M, Han M, Hashimoto K. Comparison of ketamine, 7,8-dihydroxyflavone, and ANA-12 antidepressant effects in the social defeat stress model of depression. Psychopharmacology (Berl). 2015;232(23):4325–35.
69. Shirayama Y, Yang C, Zhang JC, Ren Q, Yao W, Hashimoto K. Alterations in brain-derived neurotrophic factor (BDNF) and its precursor proBDNF in the brain regions of a learned helplessness rat model and the antidepressant effects of a TrkB agonist and antagonist. Eur Neuropsychopharmacol. 2015;25(12):2449–58.
70. Autry AE, Adachi M, Nosyreva E, Na ES, Los MF, Cheng PF, Kavalali ET, Monteggia LM. NMDA receptor blockade at rest triggers rapid behavioural antidepressant responses. Nature. 2011;475(7354):91–5.
71. Lepack AE, Fuchikami M, Dwyer JM, Banasr M, Duman RS. BDNF release is required for the behavioral actions of ketamine. Int J Neuropsychopharmacol. 2014;18(1):pyu033.
72. Svenningsson P, Kim Y, Warner-Schmidt J, Oh YS, Greengard P. p11 and its role in depression and therapeutic responses to antidepressants. Nat Rev Neurosci. 2013;14(10):673–80.
73. Sun HL, Zhou ZQ, Zhang GF, Yang C, Wang XM, Shen JC, Hashimoto K, Yang JJ. Role of hippocampal p11 in the sustained antidepressant effect of ketamine in the chronic unpredictable mild stress model. Transl Psychiatry. 2016;6:e741.
74. Liu WX, Wang J, Xie ZM, Xu N, Zhang GF, Jia M, Zhou ZQ, Hashimoto K, Yang JJ. Regulation of glutamate transporter 1 via BDNF-TrkB signaling plays a role in the anti-apoptotic and antidepressant effects of ketamine in chronic unpredictable stress model of depression. Psychopharmacology (Berl). 2016;233(3):405–15.
75. Zanos P, Moaddel R, Morris PJ, Georgiou P, Fischell J, Elmer GI, Alkondon M, Yuan P, Pribut HJ, Singh NS, Dossou KS, Fang Y, Huang XP, Mayo CL, Wainer IW, Albuquerque EX,

Thompson SM, Thomas CJ, Zarate Jr CA, Gould TD. NMDAR inhibition-independent antidepressant actions of ketamine metabolites. Nature. 2016;533(7604):481–6.
76. Shirayama Y, Hashimoto K. Effects of a single bilateral infusion of R-ketamine in the rat brain regions of a learned helplessness model of depression. Eur Arch Psychiatry Clin Neurosci. 2016. doi:10.1007/s00406-016-0718-1.
77. Yang C, Qu Y, Abe M, Nozawa D, Chaki S, Hashimoto K. (*R*)-ketamine shows greater potency and longer lasting antidepressant effects than its metabolite (2*R*,6*R*)-hydroxynorketamine. Biol Psychiatry. 2017. doi:10.1016/j.biopsych.2016.12.020.
78. Singh JB, Fedgchin M, Daly E, Xi L, Melman C, De Bruecker G, Tadic A, Sienaert P, Wiegand F, Manji H, Drevets WC, Van Nueten L. Intravenous esketamine in adult treatment-resistant depression: a double-blind, double-randomization, placebo-controlled study. Biol Psychiatry. 2016;80(6):424–31.
79. Lapidus KA, Levitch CF, Perez AM, Brallier JW, Parides MK, Soleimani L, Feder A, Iosifescu DV, Charney DS, Murrough JW. A randomized controlled trial of intranasal ketamine in major depressive disorder. Biol Psychiatry. 2014;76(12):970–6.
80. Zhang JC, Li SX, Hashimoto K. R(−)-ketamine shows greater potency and longer lasting antidepressant effects than S(+)-ketamine. Pharmacol Biochem Behav. 2014;116:137–41.
81. Yang C, Shirayama Y, Zhang JC, Ren Q, Yao W, Ma M, Dong C, Hashimoto K. R-ketamine: a rapid-onset and sustained antidepressant without psychotomimetic side effects. Transl Psychiatry. 2015;5:e632.
82. Yang C, Hashimoto K. Rapid antidepressant effects and abuse liability of ketamine. Psychopharmacology (Berl). 2014;231(9):2041–2.
83. Hashimoto K, Kakiuchi T, Ohba H, Nishiyama S, Tsukada H. Reduction of dopamine $D_{2/3}$ receptor binding in the striatum after a single administration of esketamine, but not *R*-ketamine: a PET study in conscious monkeys. Eur Arch Psychiatry Clin Neurosci. 2016. doi:10.1007/s00406-016-0692-7.
84. Murrough JW, Perez AM, Pillmer S, Stem J, Parides MK, Aan het Rot M, Collins KA, Mathew SJ, Charney DS, Iosifescu DV. Rapid and longer-term antidepressant effects of repeated ketamine infusions in treatment-resistant major depression. Biol Psychiatry. 2013;74(4):250–6.
85. Szymkowicz SM, Finnegan N, Dale RM. A 12-month naturalistic observation of three patients receiving repeat intravenous ketamine infusions for their treatment-resistant depression. J Affect Disord. 2013;147(1–3):416–20.
86. Yang C, Han M, Zhang JC, Ren Q, Hashimoto K. Loss of parvalbumin-immunoreactivity in mouse brain regions after repeated intermittent administration of esketamine, but not R-ketamine. Psychiatry Res. 2016;239:281–3.
87. Gonzalez-Burgos G, Cho RY, Lewis DA. Alterations in cortical network oscillations and parvalbumin neurons in schizophrenia. Biol Psychiatry. 2015;77(12):1031–40.
88. Lodge DJ, Behrens MM, Grace AA. A loss of parvalbumin-containing interneurons is associated with diminished oscillatory activity in an animal model of schizophrenia. J Neurosci. 2009;29(8):2344–54.
89. Moskal JR, Burgdorf JS, Stanton PK, Kroes RA, Disterhoft JF, Burch RM, Amin Khan M. The development of rapastinel (formerly GLYX-13); a rapid acting and long lasting antidepressant. Curr Neuropharmacol. 2016. doi:10.2174/1570159X14666160321122703.
90. Preskorn S, Macaluso M, Mehra DO, Zammit G, Moskal JR, Burch RM, GLYX-13 Clinical Study Group. Randomized proof of concept trial of GLYX-13, an *N*-methyl-D-aspartate receptor glycine site partial agonist, in major depressive disorder nonresponsive to a previous antidepressant agent. J Psychiatr Pract. 2015;21(2):140–9.
91. Yang B, Zhang JC, Yao W, Yang C, Ren Q, Ma M, Chen QX, Hashimoto K. Comparison of R-ketamine and rapastinel antidepressant effects in the social defeat stress model of depression. Psychopharmacology (Berl). 2016;233(19-20):3647–57.
92. Hashimoto K. Ketamine's antidepressant action: beyond NMDA receptor inhibition. Expert Opin Ther Targets. 2016;20(11):1389–92.
93. Newport DJ, Schatzberg AF, Nemeroff CB. Whither ketamine as an antidepressant: panacea or toxin? Depress Anxiety. 2016;363(8):685–8.
94. Freedman R. Further investigation of ketamine. Am J Psychiatry. 2016;173(8):761–2.

Chapter 5
D-Aspartate, an Atypical Amino Acid with NMDA Receptor Agonist Features: Involvement in Schizophrenia

F. Errico and A. Usiello

Abstract The atypical amino acid D-aspartate is transiently present in the mammalian brain. It is abundant during embryonic phases and strongly decreases after birth, when it is catabolized by the flavoenzyme D-aspartate oxidase (DDO). Pharmacological evidence indicates that D-aspartate binds to and activates NMDA receptors (NMDARs) and occurs at extracellular level where it is released through calcium-dependent mechanism. In the last 10 years, studies on mice with non-physiological high levels of D-aspartate have revealed that this D-amino acid is able to enhance NMDAR-dependent synaptic plasticity, dendritic morphology and spatial memory during adulthood. In line with the hypothesis of a NMDAR hypofunction in the pathogenesis of schizophrenia, it has been also shown that increased D-aspartate reduces prepulse inhibition deficit induced by phencyclidine, and produces corticostriatal adaptations resembling those observed after chronic haloperidol treatment. Moreover, greater D-aspartate levels can significantly inhibit functional circuits activated by phencyclidine, and increase cortico–hippocampal connectivity networks, reported to be altered in patients with schizophrenia. Besides studies in preclinical models, it has been shown that genetic variation in *DDO* gene, predicting potential increase in D-aspartate levels in *post-mortem* prefrontal cortex, is associated with greater prefrontal gray matter and activity during working memory. Interestingly, a significant reduction of D-aspartate content has been detected in

F. Errico (✉)
Laboratory of Behavioural Neuroscience, Ceinge Biotecnologie Avanzate,
Via G. Salvatore 482, Naples 80145, Italy

Department of Molecular Medicine and Medical Biotechnology, University of Naples
"Federico II", Naples 80138, Italy
e-mail: francesco.errico@unina.it

A. Usiello (✉)
Laboratory of Behavioural Neuroscience, Ceinge Biotecnologie Avanzate,
Via G. Salvatore 482, Naples 80145, Italy

Department of Environmental, Biological and Pharmaceutical Sciences and Technologies,
The Second University of Naples (SUN), Caserta 81100, Italy
e-mail: usiello@ceinge.unina.it

K. Hashimoto (ed.), *The NMDA Receptors*, The Receptors 30,
DOI 10.1007/978-3-319-49795-2_5

the *post-mortem* brain of patients with schizophrenia, associated with increased expression of *DDO* mRNA. Based on the agonistic role of D-aspartate on NMDARs and on its abundance during prenatal life, future studies will be crucial to address the biological significance of this molecule on developmental processes controlled by NMDARs and relevant to schizophrenia.

Keywords D-Aspartate • D-Aspartate oxidase • NMDA receptor • Schizophrenia • Synaptic plasticity • Prepulse inhibition • Cognition • Phencyclidine

Abbreviations

ACSF	Artificial cerebrospinal fluid
BOLD	Blood oxygen level-dependent
CA1	Cornu ammonis area 1
CBV	Cerebral blood volume
D-Asp	D-Aspartate
DDO	D-Aspartate oxidase
D-Ser	D-Serine
E-LTP	Early phase long-term potentiation
fMRI	Functional magnetic resonance imaging
HPLC	High performance liquid chromatography
L-Asp	L-Aspartate
L-Glu	L-Glutamate
L-LTP	Late phase long-term potentiation
NMDA	*N*-methyl D-aspartate
NMDAR	*N*-methyl D-aspartate receptor
PCP	Phencyclidine
PFC	Prefrontal cortex
PPI	Prepulse inhibition

5.1 Introduction

Until a few decades ago, the presence of D-amino acids was believed to be restricted only to bacteria and low animal species [1, 2]. Accumulating results since 1986 [3] have surprisingly shown that D-forms of amino acids are evolutionary well conserved in mammals, including humans, in which they are present in several tissue types. In particular, significant amounts of free D-aspartate (D-Asp) and D-serine (D-Ser) were found in the brain where they follow a peculiar regional and temporal pattern of emergence [3–7]. Such discovery, together with the long-established identification of flavoenzymes responsible for the oxidative deamination of bicarboxylic and neutral D-amino acids [8, 9], greatly supported the hypothesis that these atypical molecules might have a defined biological role in mammals. Today, it is well established that D-Ser is a physiological endogenous co-agonist for synaptic NMDA receptors (NMDARs) at central excitatory synapses of mammalian brain [10–13] where it influences development [14], synaptic transmission and plasticity [15–21], and behaviours such as cognition, sensorimotor gating [22–25] and social interaction [26–28]. On the other side, altered levels of D-Ser seem to produce a disturbed NMDAR-dependent signaling and can be causative for several pathological conditions including schizophrenia [29–31]. Differently from D-Ser, the neurobiological role of D-Asp and its influence on glutamatergic neurotransmission is becoming clearer only in the last years. In this chapter, we will overview the main advances on D-Asp, starting from the first pharmacological and neurochemical findings until the last evidence showing a potential implication of this D-amino acid in schizophrenia-related processes.

5.2 Epigenetic Changes in D-*Aspartate Oxidase* Gene Control D-Aspartate Levels in the Mammalian Brain

Detection of D-Asp in mammalian brain homogenates has shown that this D-amino acid transiently occurs in rodents and humans, since it is abundant at developmental stages and drastically decreases after birth [3, 5–7, 32, 33]. Interestingly, high performance liquid chromatography (HPLC) analysis performed on human prefrontal cortex (PFC) unveiled that D-Asp amount at gestational week 14 exceeds the levels of the corresponding L-form [5]. In the rat brain, two different immunohistochemical studies have investigated the presence of D-Asp in the rat brain at embryonic [32] and postnatal [33] developmental phases. In the prenatal brain, D-Asp appears at embryonic day 12 (E12) in the ventrocaudal regions of the forebrain, in the midbrain and hindbrain where it is localized in the cytoplasm of neuroblasts, which have already ceased proliferative activity, but not in mitotic cells [32]. In the ventrocaudal forebrain, D-Asp appears in cell bodies of neuroblasts that migrate towards the outer layer of neural epithelium. When the migration process is completed and the layer has been established, D-Asp shifts to axons. Between E14 and E20, D-Asp occurrence increases and extends to the whole brain, including the cerebral cortex. In another work [33], D-Asp has been found in considerable amount in forebrain regions of newborn rats, including the cerebral cortex, olfactory bulbs, thalamus and hypothalamus, and in part of the midbrain. At post-natal day 2 (P2), the staining also extends caudally to the hindbrain and cerebellum. At this stage D-Asp is concentrated in zones actively involved in developmental processes, in neurons that have not yet reached their final localization. Starting from P7, D-Asp levels substantially decrease. At P28, D-Asp is visible only in restricted areas of the brain [34]. Besides neuronal cells, no evidence for D-Asp staining in glia has been so far collected [33, 34].

The transient occurrence of D-Asp and changes in its localization in the developing brain imply that the endogenous levels of this D-amino acid must be regulated by a dedicated biochemical mechanism controlling both biosynthetic and degradative processes. In support of an endogenous biosynthesis of D-Asp, data in cultured PC12 mammalian cells show that D-Asp levels increase in both cells and culture media with the duration of culture [35]. Moreover, in primary rat embryonic neurons, [^{14}C]-L-aspartate (L-Asp) added to the culture media, is converted in [^{14}C]-D-Asp in a pyridoxal phosphate-dependent manner [33]. However, so far there is no general agreement on the identification of the D-aspartate-synthesizing enzyme [36–39].

If the mechanism of D-Asp biosynthesis is still controversial in mammals, the existence of an enzyme catabolising free D-Asp, D-aspartate oxidase (DDO, EC 1.4.3.1), has long been established [9]. DDO is a flavin adenine dinucleotide-containing enzyme [40] which oxidizes D-Asp in presence of H_2O and O_2, producing α-oxaloacetate, H_2O_2 and NH_4^+ ions [41]. DDO also oxidizes other bicarboxylic D-amino acids *in vitro*, such as D-glutamate and *N*-methyl D-aspartate (NMDA), while it is inactive towards basic and neutral D-amino acids, including D-Ser [42], that are degraded by the D-amino acid oxidase (DAAO, EC 1.4.3.3), a flavoenzyme homologous to DDO [43–45]. The DDO sequence possesses a functional C-terminal tripeptide for the targeting to peroxisomes [42, 46], where this enzyme is supposed to oxidize D-Asp and to safely release its toxic catabolite, H_2O_2 [47, 48]. In the brain, DDO activity strongly increases from birth until 6 weeks of life [40], and is predominantly localized in neuronal population [49] with an expression pattern reciprocal to the localization of its physiological substrate, D-Asp [34]. The onset of DDO activity after birth and its progressive increase imply a control of this enzyme on the postnatal levels of D-Asp. However, it has been unclear for long time whether the time-dependent increase of DDO activity during postnatal life correlates with *Ddo* gene transcription. In this regard, a very recent finding in mice indicates that the gradual decrease of D-Asp content (Fig. 5.1a, b), in a time-window between E15 and P60, is accompanied by complementary increased transcription of *Ddo* gene [50] (Fig. 5.1c, d), thus matching with the postnatal enhancement of DDO activity [40]. Interestingly, the postnatal increase in *Ddo* mRNA levels is reflected by progressive demethylation in the CpG sites of *Ddo* surrounding the transcription start site (8 CpG residues from −363 to +113 bp) (Fig. 5.1e, f). This observation seems to have functional impact on *Ddo* gene transcription since treatment with the demethylating agent azacitidine substantially triggers *Ddo* transcription in primary neuronal cultures from embryonic cortex that, physiologically, do not yet express this gene [50].

5.3 Pharmacological Features of D-Aspartate

In the 80s, neuropharmacological studies aimed at finding novel agonists or antagonists for ionotropic glutamate (L-Glu) receptors revealed that D-Asp binds to the L-Glu site of NMDA receptors (NMDARs) [51–55]. In line with this observation, recent evidence revealed that local applications of D-Asp on adult mouse brain

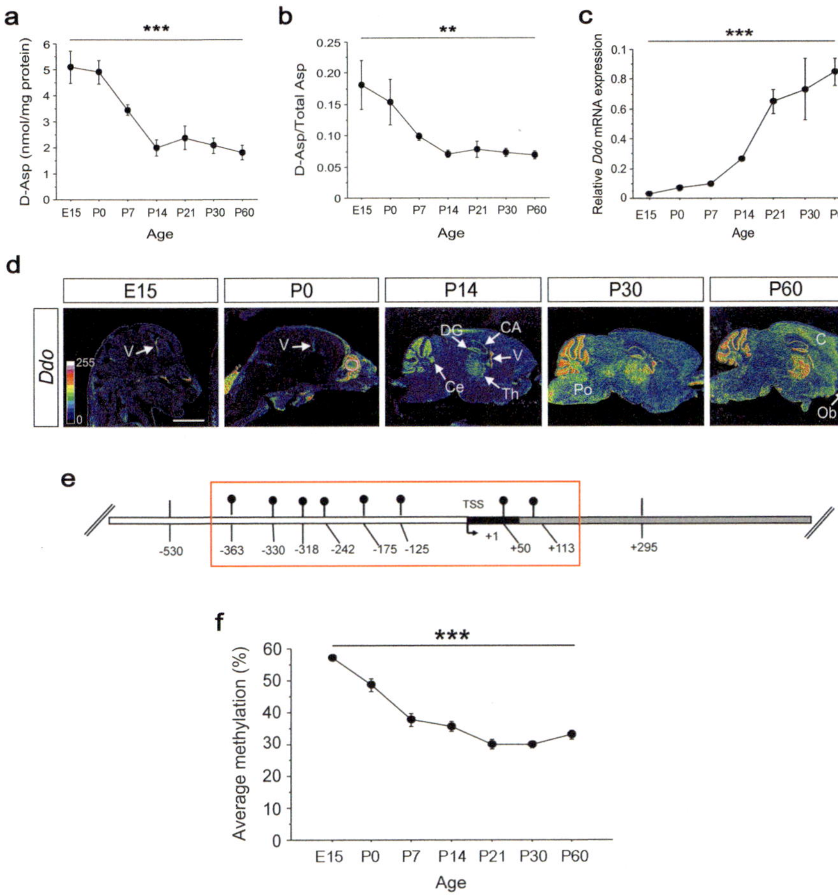

Fig. 5.1 Ontogenetic decrease of D-aspartate levels in mice is associated with demethylation within *Ddo* gene. (**a**) Average content of D-aspartate (D-Asp) and (**b**) D-aspartate/total aspartate ratio in total brain homogenates of C57BL/6J mice at different embryonic (E) and postnatal (P) days (E15, P0, P7, P14, P21, P30 and P60). The amount of D-aspartate in the brain homogenates is detected by HPLC and normalized by the total protein content of each sample. (**c**) Analysis of *Ddo* mRNA expression performed by quantitative reverse transcription PCR (qRT)-PCR on whole brain homogenates of C57BL/6J mice at different ages. Quantity means of transcript are normalized to *β-actin* housekeeping gene. (**d**) Representative pseudocolor autoradiographs from sagittal sections throughout the mouse brain at E15, P0, P14, P30 and P60 showing *Ddo* mRNA expression pattern. The relationship between autoradiographic signal intensity (from 0 to 255) and the pseudocolor images is scaled by pseudocolor bars (from *blue* to *white*). Scale bar: 750 μm. *V* ependymal cell layer of the ventricle, *CA Cornu Ammonis* area, *Ce* cerebellum, *DG* dentate gyrus, *Th* thalamus, *Po* pons, *C* cortex, *Ob* olfactory bulb. (**e**) Structure of the region surrounding the transcription start site (TSS, +1, indicated by an *arrow*) of the mouse *Ddo* gene. The putative regulatory upstream region (*white box*), exon 1 (*black*) and first intron (grey box) are indicated. Position of CpG sites is indicated as relative to TSS. The eight CpG sites analyzed are enclosed in a *red box*. (**f**) Average methylation degree of the eight CpG sites analyzed, at the indicated developmental stages. $^{**}P < 0.01$, $^{***}P < 0.0001$, one-way ANOVA with repeated measures. All the values are expressed as the mean ± SEM. Modified from Punzo et al. [50]

slices trigger inward NMDAR-dependent currents both in cornu ammonis 1 (CA1) pyramidal neurons of the hippocampus, and in the GABAergic striatal medium spiny neurons since they are antagonized by competitive and non-competitive NMDAR blockers, like D-AP5 and MK801, respectively [56, 57]. In particular, in the CA1 area of mouse hippocampus, D-Asp-induced currents are inhibited by selective antagonists for GluN2A, 2B and 2C-D subunits (NVP, Ro 25-6981 and cis-PPDA, respectively), indicating that the activation of NMDARs by D-Asp is mediated by interaction with the L-Glu site of each of these GluN2 subunit. Interestingly, residual D-Asp-dependent currents still persist after the simultaneous perfusion of selective GluN2A, 2B and 2C-D antagonists or after the application of high concentrations of D-AP5 or MK-801 [56–59], and are blocked only by switching the normal artificial cerebrospinal fluid (ACSF) solution to one not containing Ca^{2+} ions [59]. This effect suggests that D-Asp may also trigger NMDAR-independent currents. In this respect, it has been previously shown that D-Asp can inhibit kainate-induced AMPAR currents in acutely isolated rat hippocampal neurons [60] or activate mGlu5 receptors, coupled to polyphosphoinositide hydrolysis, in neonate rat hippocampal and cortical slices [61]. Likewise, a recent study performed on dopamine neurons of the *substantia nigra pars compacta* of mice has shown that currents produced by D-Asp are mainly dependent by NMDARs but a smaller component is also mediated by ionotropic AMPA receptors and metabotropic Glu1/5 receptors [62]. Finally, as expected in case of NMDAR activation, inward currents triggered by D-Asp are associated with a transient increase of intracellular Ca^{2+} in hippocampal pyramidal neurons [59].

If D-Asp is actually able to mediate neuronal communication in the brain via NMDAR-dependent transmission, then there should be a mechanism to allow also the extracellular release and the subsequent reuptake of this D-amino acid. Different works using mammalian tissue slices, cells or synaptosomal preparations have suggested that D-Asp can be actively stored in secretory organelles [63] and released through vesicular Ca^{2+}-mediated exocytotic processes [33, 63–66]. In addition to Ca^{2+}-dependent process, spontaneous release [33, 67, 68] and L-Glu transporter exchange [69, 70] have also been suggested as mechanisms responsible for D-Asp efflux.

On the other side, intracellular uptake of D-Asp has been hypothesized to occur through L-Glu/L-Asp transporter, a carrier system that utilizes the Na^{+}/K^{+} electrochemical gradient to move excitatory amino acids against their concentration gradient. Indeed, this carrier system is stereoselective for L-Glu but, interestingly, recognizes and transports both L- and D-Asp with the same efficiency [71]. Experimental approaches using [^{3}H]-D-Asp autoradiography [72] or immunostaining with D-Asp antibody [73] have demonstrated that D-Asp, preloaded on rat hippocampal slices, shows a laminar distribution identical to L-Glu, corresponding to the terminal areas of the main excitatory fiber pathways of the hippocampus [72, 73]. The reuptake of D-Asp has been observed at both nerve terminals of asymmetrical synapses and glia, probably as the result of regional and subtype heterogeneity of the transporter system [73, 74].

Besides the *in vitro* evidence reported above, the prerequisite to sustain that D-Asp is physiologically involved in NMDAR-related neurotransmission is the demonstration that this D-amino acid occurs *in vivo* at extracellular level, where it can stimulate its target receptors. This evidence has recently turned out from a microdialysis study demonstrating that D-Asp is present at nanomolar concentrations in the extracellular space of the PFC of freely moving mice [50] (Fig. 5.2a). Interestingly, when dialysate fraction is collected in a Ca^{2+}-free ACSF, extracellular D-Asp levels become undetectable, suggesting that D-Asp is released in a Ca^{2+}-dependent manner and that the pre-existing D-Asp has been efficiently removed from the extracellular space [50] (Fig. 5.2a). Moreover, the lack of DDO in *Ddo* knockout mice (*Ddo*$^{-/-}$) [75] leads to the concomitant increase of extracellular (Fig. 5.2a) and total D-Asp content (Fig. 5.2b, c) in the PFC, indicating that impaired catabolism of D-Asp affects the extracellular release of this D-amino acid [50]. Like *Ddo* gene ablation, also chronic and acute treatments with D-Asp in freely moving C57BL/6J mice are able to produce a substantial increase of D-Asp levels, both in cortical homogenates and dialysates (Errico F. and Usiello A., personal communication). The discovery of augmented extracellular D-Asp content suggests that exogenous D-Asp can efficiently cross the blood brain barrier and reach the brain parenchyma. Interestingly, both chronic and acute administrations of D-Asp are able to evoke also cortical L-Glu efflux, most likely through the stimulation of presynaptic NMDA, AMPA and mGlu5 receptors (Errico F. and Usiello A., personal communication). This evidence suggests that D-Asp can influence glutamatergic neurotransmission not only by direct stimulation of NMDARs (and, to a lesser efficacy, of non-NMDA receptors) but also indirectly, by triggering L-Glu efflux.

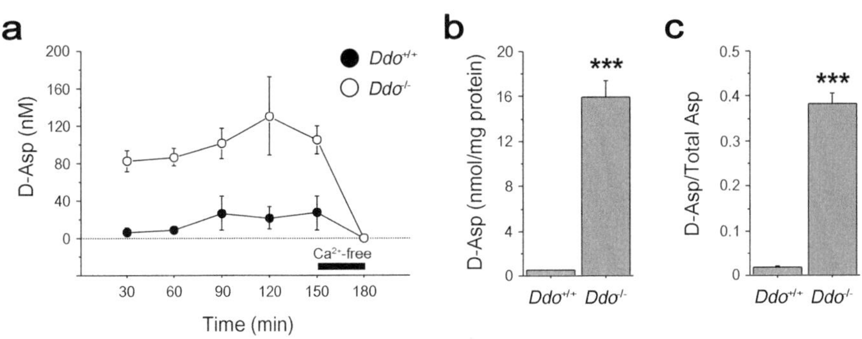

Fig. 5.2 D-aspartate is present at extracellular level in the mouse prefrontal cortex and its content increases in *Ddo*$^{-/-}$ animals. (**a**) Average extracellular concentration of D-aspartate measured by HPLC in dialysates collected from the prefrontal cortex of freely moving *Ddo*$^{+/+}$ and *Ddo*$^{-/-}$ mice in microdialysis experiments, at different time-points (30–180 min). Last fraction of dialysates (150–180 min) was collected in a Ca^{2+}-free ACSF. The reset of extracellular D-aspartate levels in Ca^{2+}-free ACSF suggests that this D-amino acid is released in a Ca^{2+}-dependent manner and that it is efficiently removed from the extracellular space. (**b**) D-aspartate (D-Asp) content and (**c**) D-aspartate/total aspartate ratio in prefrontal cortex homogenates of *Ddo*$^{+/+}$ and *Ddo*$^{-/-}$ mice, measured by HPLC and normalized by the total protein content in the tissue samples. $^{***}P < 0.0001$, compared with *Ddo*$^{+/+}$ mice (Student's t test). All the values are expressed as the mean ± SEM. Modified from Punzo et al. [50]

5.4 Abnormal Increase of D-Aspartate in the Brain Affects Synaptic Plasticity, Dendritic Morphology and Cognition

In agreement with the ability of D-Asp to stimulate glutamatergic transmission, increased levels of D-Asp in both adult *Ddo*$^{-/-}$ and chronically D-Asp-treated mice are associated with enhanced NMDAR-dependent early phase long-term potentiation (E-LTP) in the hippocampal CA1 area of mice [56, 58, 59]. Experiments of intermittent oral administration of D-Asp to C57BL/6J animals indicate that the magnitude of NMDAR-dependent E-LTP in the hippocampus is regulated by changes in the brain levels of this D-amino acid. Indeed, 3 weeks of interruption after chronic D-Asp administration is able to wash out the excess of D-Asp and, in turn, to normalize E-LTP amplitude. Interestingly, 1-month treatment with D-Asp, after 3-week withdrawal, re-establishes synaptic plasticity at previously potentiated levels [58]. Elevated D-Asp levels can influence also long-lasting forms of hippocampal synaptic plasticity. Indeed, paradigm for E-LTP induction, that causes a decaying LTP in wild-type slices, is sufficient to induce stable late phase LTP (L-LTP) in slices from *Ddo*$^{-/-}$ and D-Asp-treated animals [76]. In both animal models, D-Asp-dependent L-LTP is insensible to rapamycin but is fully prevented by cytochalasin D administration [76]. The D-Asp-dependent enhancement in LTP may occur through direct activation of NMDARs by D-Asp and/or by indirect influence of D-Asp on the release of endogenous L-Glu following tetanic stimulation. In line with a role for D-Asp on NMDAR-dependent transmission, it has been shown that mice chronically treated with D-Asp also revealed increased frequency in NMDAR-mediated miniature excitatory post-synaptic currents recorded in pyramidal neurons of the medial PFC layer II/III [76]. The enhancement of glutamatergic transmission in D-Asp-treated mice is mirrored by greater basal metabolic activity in fronto-hippocampal areas, as assessed with basal cerebral blood volume (CBV)-weighted functional magnetic resonance imaging (fMRI) [76].

Moreover, in line with enhanced NMDAR-dependent transmission and facilitated induction of late-phase synaptic plasticity, elevation of D-Asp levels is also associated with modifications in neuronal cytoarchitecture since both *Ddo*$^{-/-}$ and D-Asp-treated mice display increased dendritic length and spine density in pyramidal neurons of the PFC and hippocampus [76]. Coherently with the D-Asp-dependent potentiation of structural and functional synaptic plasticity, increased levels of D-Asp are also associated with improved spatial cognitive abilities of young-adult *Ddo*$^{-/-}$ and D-Asp-treated mice, when tested in NMDAR-dependent task involving hippocampal circuits, such as the hidden-platform version of the Morris water maze [57–59].

5.5 D-Aspartate Affects NMDA Receptor-Dependent Processes Relevant to Schizophrenia

The hypothesis of developmental NMDAR hypofunction in the pathogenesis of schizophrenia is today supported by a large body of evidence [77–79]. In this regard, several studies have suggested that reduced levels of the co-agonist D-Ser may result in disturbed NMDAR neurotransmission and be thus involved in the pathophysiology of this psychiatric disorder [29, 30, 79]. Based on this evidence, D-Ser and other molecules targeting the glycine modulatory site of NMDAR have been proposed as emerging therapeutic targets in schizophrenia [29, 31, 80–82]. In the light of the similar influence played by D-Asp on NMDAR-dependent transmission and of its neurodevelopmental occurrence, this D-amino acid could be another potential endogenous target of relevance to NMDAR-related processes involved in schizophrenia. To evaluate *in vivo* this hypothesis, mouse models with increased levels of D-Asp have been tested in the prepulse inhibition (PPI) paradigm, a cross-species operational measure of sensorimotor gating regarded today as an endophenotypic trait of schizophrenia [83, 84]. Results showed that chronic exposure to higher D-Asp levels does not affect the basal sensorimotor filtering of adult $Ddo^{-/-}$ and D-Asp-treated mice but substantially reduces the inhibitory deficits induced in these mice by acute treatment with psychotomimetic drugs like amphetamine and MK801 [57]. In support of a protective effect of increased D-Asp levels against psychotic-like deficits, recent data has shown significantly reduced motor hyperactivity (Fig. 5.3a) and PPI disruption (Fig. 5.3b) in $Ddo^{-/-}$ mice acutely treated with phencyclidine (PCP) [85]. Interestingly, behavioural resiliency is accompanied by reduced functional circuits activation in cortico-limbo-thalamic regions of PCP-treated $Ddo^{-/-}$ animals, as measured by fMRI [85] (Fig. 5.3c–e). Another attractive translational feature found in $Ddo^{-/-}$ mice is the greater connectivity found in their cortico-hippocampal circuitries, as opposed to substantial brain dysconnectivity observed in both animal models of schizophrenia and patients [86–88]. Consistent with enhanced glutamatergic synaptic strength produced by elevated D-Asp levels, both $Ddo^{-/-}$ mutant and D-Asp-treated mice display inhibition of long-term depression at corticostriatal synapses [57], an effect described under conditions of enhanced NMDAR signaling and after chronic treatment with the antipsychotic haloperidol [89].

Recent unpublished data evidenced that D-Asp relevance in schizophrenia could be potentially extended also to the mechanism of action of second-generation antipsychotics. Indeed, HPLC detections performed on PFC dialysates and homogenates revealed that chronic administration of olanzapine, but not clozapine, triggers a significant increase of D-Asp levels, respectively. Interestingly, such increase is mirrored by the ability of olanzapine, but not clozapine, to inhibit *in vitro* the activity of the human and murine recombinant DDO enzymes. This would suggest that olanzapine could influence the extracellular release of D-Asp by controlling the intracellular activity of DDO. The same study also revealed that D-Asp and olanzapine share the ability to promote the cortical release of L-Glu. Therefore, it could be hypothesized that olanzapine, besides affecting dopaminergic and serotonergic signalling [90], may also enhance glutamatergic neurotransmission through the modulation of D-Asp catabolism (Errico F. and Usiello A., personal communication).

Overall, the results collected in preclinical animal models provide an encouraging background to evaluate whether also in humans the state of D-Asp metabolism may have an impact on phenotypes relevant to schizophrenia. In this regard, a recent work has detected the levels of endogenous free D-Asp and NMDA in a small cohort of *post-mortem* PFC samples of patients with schizophrenia revealing a consistent reduction of

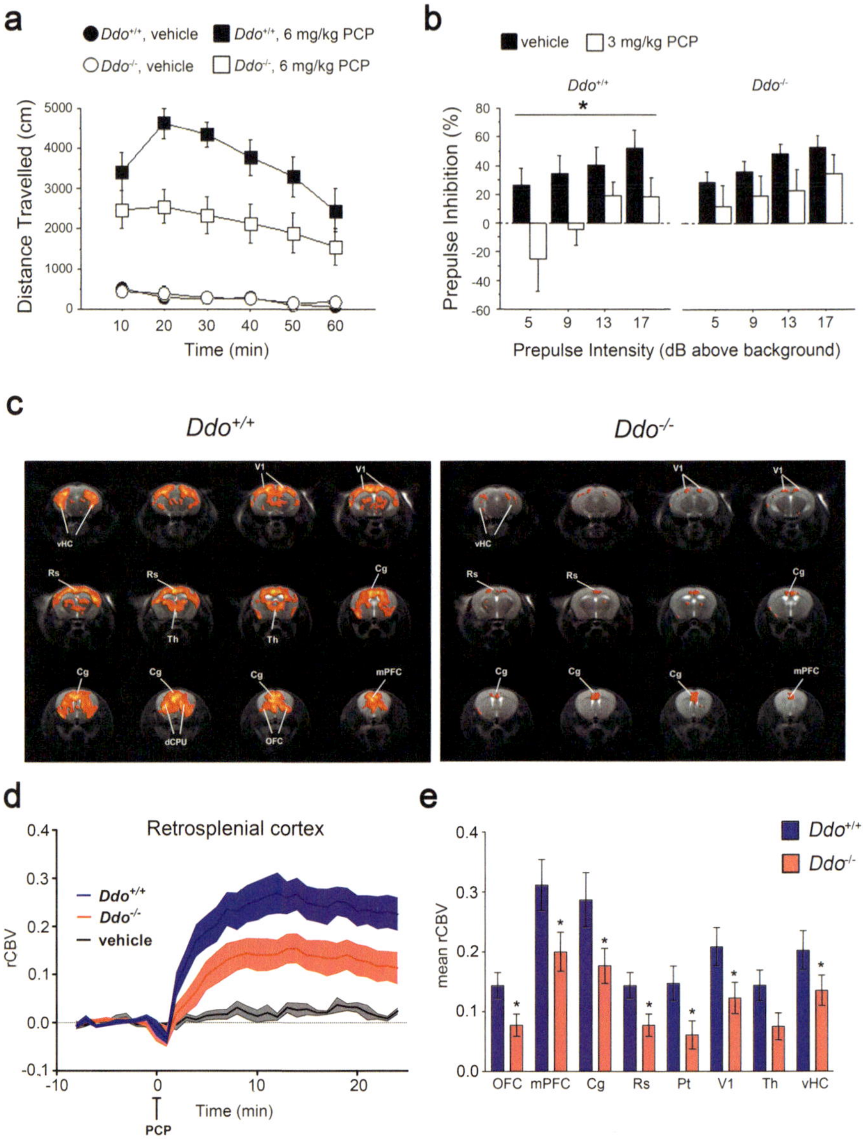

Fig. 5.3 Reduced PCP-mediated behavioural and functional responses in *Ddo*$^{-/-}$ mice. (**a**) Motor activity induced by 6 mg/kg PCP is significantly reduced in *Ddo*$^{-/-}$ mice, compared to *Ddo*$^{+/+}$ littermates. Locomotion was expressed as distance travelled, measured in cm every 10 min over a 60-min session. (**b**) Prepulse inhibition (PPI) deficits induced by 3 mg/kg PCP are attenuated in *Ddo*$^{-/-}$ mice, compared to *Ddo*$^{+/+}$ animals. Percentage of the PPI was used as dependent variable and measured at different prepulse intensities (dB above 65-dB background level). $^{*}P < 0.05$, compared with vehicle control groups (two-way ANOVA with prepulse intensity as repeated measures). (**c–e**) PCP-induced functional magnetic resonance imaging (MRI) response in *Ddo*$^{-/-}$ mice. In *Ddo*$^{+/+}$ mice, PCP elicits robust and sustained cortico-limbo-thalamic fMRI activation. This effect is strongly attenuated in *Ddo*$^{-/-}$ mice. *Red/yellow* indicates significant fMRI (relative cerebral blood volume, rCBV) response to PCP (1 mg/kg, intra-artery) with respect to vehicle (saline; 3.1 < *Z* score < 6, cluster correction threshold pc = 0.001). $^{*}P < 0.05$, Student's t test. *Cg* cingulate cortex, *dCPU* dorsal caudate putamen, *mPFC* medial prefrontal cortex, *OFC* orbito-frontal cortex, *Rs* retrosplenial cortex, *Th* thalamus, *vHc* ventral hippocampus, *V1*: visual cortex. All values are expressed as the mean ± SEM. Modified from Errico et al. [85]

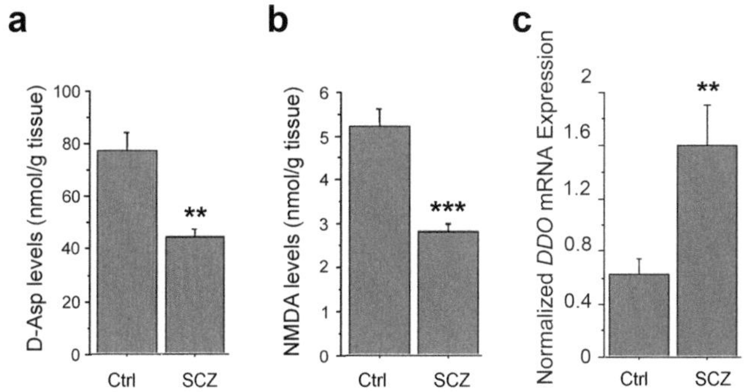

Fig. 5.4 Decreased levels of D-aspartate and NMDA in the *post-mortem* prefrontal cortex of patients with schizophrenia are mirrored by increased expression of *DDO* mRNA. (**a**) D-aspartate (D-Asp) and (**b**) *N*-methyl D-aspartate (NMDA) were measured by HPLC in the prefrontal cortex of control and schizophrenia subjects. $^{**}P < 0.01$, $^{***}P < 0.0001$, compared to control group (ANOVA). Modified from Errico et al. [91]. (**c**) Analysis of *DDO* mRNA expression was performed by quantitative reverse transcription (qRT)-PCR in the prefrontal cortex of control and schizophrenia subjects, and normalized to the geometric mean of three housekeeping genes (*β-actin*, *glyceraldehyde-3-phosphate dehydrogenase* and *cyclophilin*). $^{**}P < 0.01$, compared to control group (ANCOVA). Modified from Errico et al. [85]. Values are expressed as the mean ± SEM

both D-amino acids levels [91] (Fig. 5.4a, b), paralleled by increased *DDO* mRNA expression [85] (Fig. 5.4c). Interestingly, the decline of endogenous D-Asp and NMDA correlates with a selective reduction in the levels of the NMDAR subunits GluN1, GluN2A and GluN2B in the prefrontal cortex of schizophrenia-affected patients [91]. Another work has recently examined the association of *DDO* gene variants with a series of complex prefrontal phenotypes [76]. Analysis of data from 268 brains of non-psychiatric individuals obtained from the *post-mortem* collection bank, Braincloud (http://braincloud.jhmi.edu) [92], has evidenced that the C allele of rs3757351 is significantly associated with reduced expression of *DDO* mRNA, when compared to the T allele (Fig. 5.5a). This result predicts a potential increase of endogenous D-Asp levels in this brain region. Then, to evaluate the functional effect of this single nucleotide polymorphism, healthy individuals were subjected to voxel-based morphometry (n = 152) and to blood oxygen level-dependent (BOLD) fMRI prefrontal activity during performance of the 1- and 2-Back working memory task (n = 143). The results revealed that subjects with the C allele also display augmented prefrontal gray matter volume (Fig. 5.5b) and greater prefrontal activity (Fig. 5.5c), when compared to individuals with the T allele.

5.6 D-Aspartate Oxidase Plays a Neuroprotective Role Against Toxic Effects Produced by Prolonged D-Aspartate Exposure

We have so far brought evidence in support of ameliorative effects of D-Asp on a series of processes likely dependent on NMDAR activation. Mouse models with increased levels of D-Asp display, indeed, enhanced functional and structural

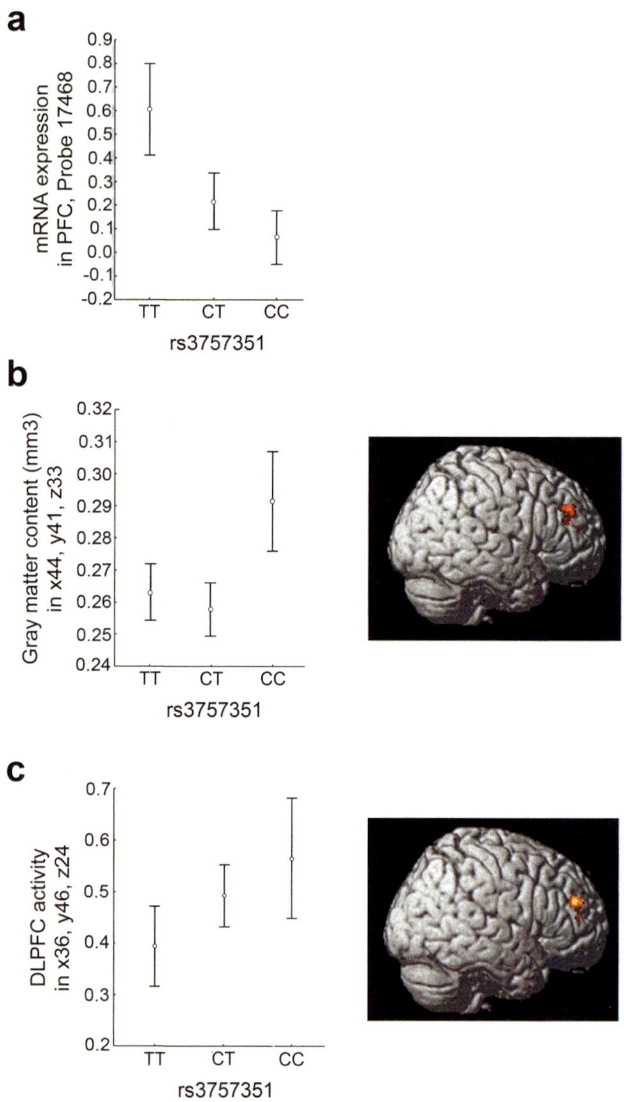

Fig. 5.5 Association of *DDO* rs3757351 with prefrontal phenotypes in humans. (**a**) Association of *DDO* rs3757351 with *DDO* mRNA expression levels in *post-mortem* prefrontal cortex (n = 268). Graph depicts normalized $\log_2$ ratios (sample/reference). These data were obtained from the largest *post-mortem* collection publicly available (BRAINCLOUD, courtesy of the Lieber Institute for Brain Development, Baltimore, USA36, http://braincloud.jhmi.edu). (**b**) Association of *DDO* rs3757351 with prefrontal gray matter volume in caucasian healthy subjects (n = 159). *Left panel*: three-dimensional rendering of the prefrontal cluster associated with a main effect of rs3757351. Image thresholded at $P < 0.005$, non-stationary cluster extend corrected. *Right panel*: graph showing mean ± 0.95 confidence intervals of gray matter content extracted from the cluster depicted in the *left panel*. (**c**) Association of *DDO* rs3757351 with prefrontal BOLD response during working memory in caucasian healthy subjects (n = 143). *Left panel*: three-dimensional rendering of the prefrontal cluster associated with a main effect of rs3757351. Image thresholded at $P < 0.05$, familywise error corrected. *Right panel*: graph showing mean ± 0.95 confidence intervals of parameter estimated extracted from the cluster depicted in the *left panel*. Modified from Errico et al. [76]

synaptic plasticity, accompanied to potentiated cortico-hippocampal connectivity, improved spatial memory, and resiliency against sensorimotor deficits and brain circuits activation induced by PCP. Taken together, these phenotypes are in line with the view that stimulation of NMDARs, within a certain limit, is crucial for promoting synaptic strength, connectivity and the formation of learning and memory [93]. By contrast, it is well known that exaggerated, chronic activation of NMDARs can lead to neuronal death and be, therefore, harmful for brain functioning [93, 94]. Given the ability of D-Asp to act as an endogenous NMDAR agonist and to amplify the L-Glu release [95], excessive D-Asp content produces neurotoxic effects during brain aging. In fact, while increased levels of endogenous D-Asp enhance the NMDAR-dependent E-LTP in 4/5-month-old $Ddo^{-/-}$ mice, the persistent up-regulation of this D-amino acid accelerates the age-related decay of synaptic plasticity in 9/10- and, even more robustly, in 13/14-month-old animals [59]. Consistently, spatial memory improvement found in 4/5-month-old $Ddo^{-/-}$ mice turns into drastic cognitive worsening at 13–14 months of age [59]. These observations match with reduced synaptic fraction of hippocampal GluN1 and GluN2B subunits observed in $Ddo^{-/-}$ mice [95]. In line with synaptic plasticity deterioration and memory deficits, two recent works have revealed precocious caspases activation and cell death in the hippocampus and PFC of $Ddo^{-/-}$ mice, accompanied to the appearance of reactive astrocytes and dystrophic microglia [50, 95]. Likewise, early expression of caspase-3 and increased cell death, together with substantial age-dependent changes in microglia morphology, have been also found in the *substantia nigra pars compacta* of $Ddo^{-/-}$ brains [50]. Like constitutive elevation of D-Asp content in $Ddo^{-/-}$ mice, long-term treatment with D-Asp for 12 months to C57BL/6J mice is able to significantly reduce E-LTP at CA1 synapses, compared to non-treated mice [58]. Interestingly, the interruption of D-Asp administration for 3 weeks, after 12-month continuous treatment, can restore hippocampal synaptic plasticity at control levels [58].

Overall, these data highlight the importance of DDO as a detoxifying enzyme able to control potential precocious NMDAR-mediated neurodegenerative events produced by excessive D-Asp stimulation. Given its ability to reduce schizophrenia-like phenotypes induced by PCP and to potentiate synaptic plasticity, dendritic structure and brain connectivity, it is possible to hypothesize the utilization of D-Asp in clinical trials, as add-on to antipsychotic treatment, to test the potential beneficial effect of this D-amino acid on schizophrenia symptoms. However, the neurotoxic effects of prolonged D-Asp exposure must be taken into account to develop an appropriate administration protocol and consider potential side effects associated to this NMDAR agonist.

5.7 Conclusions

Thanks to the development and the study of animal models with increased levels of D-Asp, it is now clear that this molecule is not just a vestigial residue inherited from primordial organisms but it can be fully deemed as an endogenous agonist of

NMDARs, influencing higher brain functions in mammals. Nevertheless, it should be remarked that the results so far collected have been obtained in mice in which the levels of D-Asp are maintained forcedly high at adulthood, when they are physiologically very low. Therefore, the generation of novel animal models with reduced levels of D-Asp since embryonic phase may help to explain the reason of the massive presence of D-Asp in the developing brain and, in turn, to clarify the potential involvement of D-Asp in schizophrenia. Indeed, it is consolidated the idea that the aetiology of this mental illness involves pathological processes, caused by both genetic and environmental factors, that begin in utero and develop until adolescence or young adulthood when they lead to the emergence of symptoms [96–98]. In this light, a putative reduction in D-Asp content during embryonic phases may have substantial clinical relevance since it could affect key neurodevelopmental processes like neurogenesis, survival, migration and generation of neuronal circuitry which are regulated by NMDARs [99–101]. Future studies will help to clarify this crucial topic.

References

1. Corrigan JJ. D-amino acids in animals. Science. 1969;164(876):142–9.
2. D'Aniello A, Giuditta A. Identification of D-aspartic acid in the brain of *Octopus vulgaris* Lam. J Neurochem. 1977;29(6):1053–7.
3. Dunlop DS, Neidle A, McHale D, Dunlop DM, Lajtha A. The presence of free D-aspartic acid in rodents and man. Biochem Biophys Res Commun. 1986;141(1):27–32.
4. Hamase K, Homma H, Takigawa Y, Fukushima T, Santa T, Imai K. Regional distribution and postnatal changes of D-amino acids in rat brain. Biochim Biophys Acta. 1997;1334(2–3):214–22.
5. Hashimoto A, Kumashiro S, Nishikawa T, Oka T, Takahashi K, Mito T, et al. Embryonic development and postnatal changes in free D-aspartate and D-serine in the human prefrontal cortex. J Neurochem. 1993;61(1):348–51.
6. Neidle A, Dunlop DS. Developmental changes in free D-aspartic acid in the chicken embryo and in the neonatal rat. Life Sci. 1990;46(21):1517–22.
7. Hashimoto A, Oka T, Nishikawa T. Anatomical distribution and postnatal changes in endogenous free D-aspartate and D-serine in rat brain and periphery. Eur J Neurosci. 1995;7(8):1657–63.
8. Krebs HA. Metabolism of amino-acids: Deamination of amino-acids. Biochem J. 1935;29(7):1620–44.
9. Still JL, Buell MV, et al. Studies on the cyclophorase system; D-aspartic oxidase. J Biol Chem. 1949;179(2):831–7.
10. Martineau M, Baux G, Mothet JP. D-serine signalling in the brain: friend and foe. Trends Neurosci. 2006;29(8):481–91.
11. Wolosker H, Radzishevsky I. The serine shuttle between glia and neurons: implications for neurotransmission and neurodegeneration. Biochem Soc Trans. 2013;41(6):1546–50.
12. Billard JM. D-Amino acids in brain neurotransmission and synaptic plasticity. Amino Acids. 2012;43(5):1851–60.
13. Van Horn MR, Sild M, Ruthazer ES. D-serine as a gliotransmitter and its roles in brain development and disease. Front Cell Neurosci. 2013;7:39.
14. Kim PM, Aizawa H, Kim PS, Huang AS, Wickramasinghe SR, Kashani AH, et al. Serine racemase: activation by glutamate neurotransmission via glutamate receptor interacting protein and mediation of neuronal migration. Proc Natl Acad Sci U S A. 2005;102(6):2105–10.

15. Mothet JP, Parent AT, Wolosker H, Brady Jr RO, Linden DJ, Ferris CD, et al. D-serine is an endogenous ligand for the glycine site of the N-methyl-D-aspartate receptor. Proc Natl Acad Sci U S A. 2000;97(9):4926–31.
16. Papouin T, Ladepeche L, Ruel J, Sacchi S, Labasque M, Hanini M, et al. Synaptic and extrasynaptic NMDA receptors are gated by different endogenous coagonists. Cell. 2012;150(3):633–46.
17. Rosenberg D, Artoul S, Segal AC, Kolodney G, Radzishevsky I, Dikopoltsev E, et al. Neuronal D-serine and glycine release via the Asc-1 transporter regulates NMDA receptor-dependent synaptic activity. J Neurosci. 2013;33(8):3533–44.
18. Li Y, Sacchi S, Pollegioni L, Basu AC, Coyle JT, Bolshakov VY. Identity of endogenous NMDAR glycine site agonist in amygdala is determined by synaptic activity level. Nat Commun. 2013;4:1760.
19. Yang Y, Ge W, Chen Y, Zhang Z, Shen W, Wu C, et al. Contribution of astrocytes to hippocampal long-term potentiation through release of D-serine. Proc Natl Acad Sci U S A. 2003;100(25):15194–9.
20. Fossat P, Turpin FR, Sacchi S, Dulong J, Shi T, Rivet JM, et al. Glial D-serine gates NMDA receptors at excitatory synapses in prefrontal cortex. Cereb Cortex. 2012;22(3):595–606.
21. Meunier CN, Dallerac G, Le Roux N, Sacchi S, Levasseur G, Amar M, et al. D-serine and glycine differentially control neurotransmission during visual cortex critical period. PLoS One. 2016;11(3):e0151233.
22. Hagiwara H, Iyo M, Hashimoto K. Neonatal disruption of serine racemase causes schizophrenia-like behavioral abnormalities in adulthood: clinical rescue by d-serine. PLoS One. 2013;8(4):e62438.
23. Hashimoto K, Fujita Y, Horio M, Kunitachi S, Iyo M, Ferraris D, et al. Co-administration of a D-amino acid oxidase inhibitor potentiates the efficacy of D-serine in attenuating prepulse inhibition deficits after administration of dizocilpine. Biol Psychiatry. 2009;65(12):1103–6.
24. Kanahara N, Shimizu E, Ohgake S, Fujita Y, Kohno M, Hashimoto T, et al. Glycine and D: -serine, but not D: -cycloserine, attenuate prepulse inhibition deficits induced by NMDA receptor antagonist MK-801. Psychopharmacology. 2008;198(3):363–74.
25. Matsuura A, Fujita Y, Iyo M, Hashimoto K. Effects of sodium benzoate on pre-pulse inhibition deficits and hyperlocomotion in mice after administration of phencyclidine. Acta Neuropsychiatr. 2015;27(3):159–67.
26. Labrie V, Duffy S, Wang W, Barger SW, Baker GB, Roder JC. Genetic inactivation of D-amino acid oxidase enhances extinction and reversal learning in mice. Learn Mem. 2009;16(1):28–37.
27. DeVito LM, Balu DT, Kanter BR, Lykken C, Basu AC, Coyle JT, et al. Serine racemase deletion disrupts memory for order and alters cortical dendritic morphology. Genes Brain Behav. 2011;10(2):210–22.
28. Basu AC, Tsai GE, Ma CL, Ehmsen JT, Mustafa AK, Han L, et al. Targeted disruption of serine racemase affects glutamatergic neurotransmission and behavior. Mol Psychiatry. 2009;14(7):719–27.
29. Javitt DC, Zukin SR, Heresco-Levy U, Umbricht D. Has an angel shown the way? Etiological and therapeutic implications of the PCP/NMDA model of schizophrenia. Schizophr Bull. 2012;38(5):958–66.
30. Labrie V, Wong AH, Roder JC. Contributions of the D-serine pathway to schizophrenia. Neuropharmacology. 2012;62(3):1484–503.
31. Coyle JT, Tsai G. The NMDA receptor glycine modulatory site: a therapeutic target for improving cognition and reducing negative symptoms in schizophrenia. Psychopharmacology. 2004;174(1):32–8.
32. Sakai K, Homma H, Lee JA, Fukushima T, Santa T, Tashiro K, et al. Emergence of D-aspartic acid in the differentiating neurons of the rat central nervous system. Brain Res. 1998;808(1):65–71.
33. Wolosker H, D'Aniello A, Snyder SH. D-aspartate disposition in neuronal and endocrine tissues: ontogeny, biosynthesis and release. Neuroscience. 2000;100(1):183–9.

34. Schell MJ, Cooper OB, Snyder SH. D-aspartate localizations imply neuronal and neuroendocrine roles. Proc Natl Acad Sci U S A. 1997;94(5):2013–8.
35. Long Z, Homma H, Lee JA, Fukushima T, Santa T, Iwatsubo T, et al. Biosynthesis of D-aspartate in mammalian cells. FEBS Lett. 1998;434(3):231–5.
36. Kim PM, Duan X, Huang AS, Liu CY, Ming GL, Song H, et al. Aspartate racemase, generating neuronal D-aspartate, regulates adult neurogenesis. Proc Natl Acad Sci U S A. 2010;107(7):3175–9.
37. Matsuda S, Katane M, Maeda K, Kaneko Y, Saitoh Y, Miyamoto T, et al. Biosynthesis of D-aspartate in mammals: the rat and human homologs of mouse aspartate racemase are not responsible for the biosynthesis of D-aspartate. Amino Acids. 2015;47(5):975–85.
38. Tanaka-Hayashi A, Hayashi S, Inoue R, Ito T, Konno K, Yoshida T, et al. Is D-aspartate produced by glutamic-oxaloacetic transaminase-1 like 1 (Got1l1): a putative aspartate racemase? Amino Acids. 2014;47(1):79–86.
39. Horio M, Ishima T, Fujita Y, Inoue R, Mori H, Hashimoto K. Decreased levels of free D-aspartic acid in the forebrain of serine racemase (Srr) knock-out mice. Neurochem Int. 2013;62(6):843–7.
40. Van Veldhoven PP, Brees C, Mannaerts GP. D-aspartate oxidase, a peroxisomal enzyme in liver of rat and man. Biochim Biophys Acta. 1991;1073(1):203–8.
41. D'Aniello A, Vetere A, Petrucelli L. Further study on the specificity of D-amino acid oxidase and D-aspartate oxidase and time course for complete oxidation of D-amino acids. Comp Biochem Physiol B. 1993;105(3–4):731–4.
42. Setoyama C, Miura R. Structural and functional characterization of the human brain D-aspartate oxidase. J Biochem. 1997;121(4):798–803.
43. Pollegioni L, Piubelli L, Sacchi S, Pilone MS, Molla G. Physiological functions of D-amino acid oxidases: from yeast to humans. Cell Mol Life Sci. 2007;64(11):1373–94.
44. Sacchi S, Caldinelli L, Cappelletti P, Pollegioni L, Molla G. Structure-function relationships in human D-amino acid oxidase. Amino Acids. 2012;43(5):1833–50.
45. Negri A, Ceciliani F, Tedeschi G, Simonic T, Ronchi S. The primary structure of the flavoprotein D-aspartate oxidase from beef kidney. J Biol Chem. 1992;267(17):11865–71.
46. Amery L, Brees C, Baes M, Setoyama C, Miura R, Mannaerts GP, et al. C-terminal tripeptide Ser-Asn-Leu (SNL) of human D-aspartate oxidase is a functional peroxisome-targeting signal. Biochem J. 1998;336(Pt 2):367–71.
47. Katane M, Homma H. D-aspartate oxidase: the sole catabolic enzyme acting on free D-aspartate in mammals. Chem Biodivers. 2010;7(6):1435–49.
48. Beard ME. D-aspartate oxidation by rat and bovine renal peroxisomes: an electron microscopic cytochemical study. J Histochem Cytochem. 1990;38(9):1377–81.
49. Zaar K, Kost HP, Schad A, Volkl A, Baumgart E, Fahimi HD. Cellular and subcellular distribution of D-aspartate oxidase in human and rat brain. J Comp Neurol. 2002;450(3):272–82.
50. Punzo D, Errico F, Cristino L, Sacchi S, Keller S, Belardo C, et al. Age-related changes in d-Aspartate oxidase promoter methylation control extracellular d-Aspartate levels and prevent precocious cell death during brain aging. J Neurosci. 2016;36(10):3064–78.
51. Fagg GE, Matus A. Selective association of N-methyl aspartate and quisqualate types of L-glutamate receptor with brain postsynaptic densities. Proc Natl Acad Sci U S A. 1984;81(21):6876–80.
52. Monahan JB, Michel J. Identification and characterization of an N-methyl-D-aspartate-specific L-[3H]glutamate recognition site in synaptic plasma membranes. J Neurochem. 1987;48(6):1699–708.
53. Ogita K, Yoneda Y. Disclosure by triton X-100 of NMDA-sensitive [3H] glutamate binding sites in brain synaptic membranes. Biochem Biophys Res Commun. 1988;153(2):510–7.
54. Olverman HJ, Jones AW, Mewett KN, Watkins JC. Structure/activity relations of N-methyl-D-aspartate receptor ligands as studied by their inhibition of [3H]D-2-amino-5-phosphonopentanoic acid binding in rat brain membranes. Neuroscience. 1988;26(1):17–31.

55. Ransom RW, Stec NL. Cooperative modulation of [3H]MK-801 binding to the N-methyl-D-aspartate receptor-ion channel complex by L-glutamate, glycine, and polyamines. J Neurochem. 1988;51(3):830–6.
56. Errico F, Nistico R, Palma G, Federici M, Affuso A, Brilli E, et al. Increased levels of d-aspartate in the hippocampus enhance LTP but do not facilitate cognitive flexibility. Mol Cell Neurosci. 2008;37(2):236–46.
57. Errico F, Rossi S, Napolitano F, Catuogno V, Topo E, Fisone G, et al. D-aspartate prevents corticostriatal long-term depression and attenuates schizophrenia-like symptoms induced by amphetamine and MK-801. J Neurosci. 2008;28(41):10404–14.
58. Errico F, Nistico R, Napolitano F, Mazzola C, Astone D, Pisapia T, et al. Increased D-aspartate brain content rescues hippocampal age-related synaptic plasticity deterioration of mice. Neurobiol Aging. 2011;32(12):2229–43.
59. Errico F, Nistico R, Napolitano F, Oliva AB, Romano R, Barbieri F, et al. Persistent increase of D-aspartate in D-aspartate oxidase mutant mice induces a precocious hippocampal age-dependent synaptic plasticity and spatial memory decay. Neurobiol Aging. 2011;32(11):2061–74.
60. Gong XQ, Frandsen A, Lu WY, Wan Y, Zabek RL, Pickering DS, et al. D-aspartate and NMDA, but not L-aspartate, block AMPA receptors in rat hippocampal neurons. Br J Pharmacol. 2005;145(4):449–59.
61. Molinaro G, Pietracupa S, Di Menna L, Pescatori L, Usiello A, Battaglia G, et al. D-aspartate activates mGlu receptors coupled to polyphosphoinositide hydrolysis in neonate rat brain slices. Neurosci Lett. 2010;478(3):128–30.
62. Krashia P, Ledonne A, Nobili A, Cordella A, Errico F, Usiello A, et al. Persistent elevation of D-Aspartate enhances NMDA receptor-mediated responses in mouse substantia nigra pars compacta dopamine neurons. Neuropharmacology. 2015;103:69–78.
63. Nakatsuka S, Hayashi M, Muroyama A, Otsuka M, Kozaki S, Yamada H, et al. D-Aspartate is stored in secretory granules and released through a Ca(2+)-dependent pathway in a subset of rat pheochromocytoma PC12 cells. J Biol Chem. 2001;276(28):26589–96.
64. Davies LP, Johnston GA. Uptake and release of D- and L-aspartate by rat brain slices. J Neurochem. 1976;26(5):1007–14.
65. Malthe-Sorenssen D, Skrede KK, Fonnum F. Calcium-dependent release of D-[3H]aspartate evoked by selective electrical stimulation of excitatory afferent fibres to hippocampal pyramidal cells in vitro. Neuroscience. 1979;4(9):1255–63.
66. D'Aniello S, Somorjai I, Garcia-Fernandez J, Topo E, D'Aniello A. D-Aspartic acid is a novel endogenous neurotransmitter. FASEB J. 2010;25(3):1014–27.
67. Adachi M, Koyama H, Long Z, Sekine M, Furuchi T, Imai K, et al. L-Glutamate in the extracellular space regulates endogenous D-aspartate homeostasis in rat pheochromocytoma MPT1 cells. Arch Biochem Biophys. 2004;424(1):89–96.
68. Koyama H, Adachi M, Sekine M, Katane M, Furuchi T, Homma H. Cytoplasmic localization and efflux of endogenous D-aspartate in pheochromocytoma 12 cells. Arch Biochem Biophys. 2006;446(2):131–9.
69. Anderson CM, Bridges RJ, Chamberlin AR, Shimamoto K, Yasuda-Kamatani Y, Swanson RA. Differing effects of substrate and non-substrate transport inhibitors on glutamate uptake reversal. J Neurochem. 2001;79(6):1207–16.
70. Bak LK, Schousboe A, Waagepetersen HS. Characterization of depolarization-coupled release of glutamate from cultured mouse cerebellar granule cells using DL-threo-beta-benzyloxyaspartate (DL-TBOA) to distinguish between the vesicular and cytoplasmic pools. Neurochem Int. 2003;43(4–5):417–24.
71. Palacin M, Estevez R, Bertran J, Zorzano A. Molecular biology of mammalian plasma membrane amino acid transporters. Physiol Rev. 1998;78(4):969–1054.
72. Taxt T, Storm-Mathisen J. Uptake of D-aspartate and L-glutamate in excitatory axon terminals in hippocampus: autoradiographic and biochemical comparison with gamma-

aminobutyrate and other amino acids in normal rats and in rats with lesions. Neuroscience. 1984;11(1):79–100.
73. Gundersen V, Danbolt NC, Ottersen OP, Storm-Mathisen J. Demonstration of glutamate/aspartate uptake activity in nerve endings by use of antibodies recognizing exogenous D-aspartate. Neuroscience. 1993;57(1):97–111.
74. Garthwaite G, Garthwaite J. Sites of D-[3H]aspartate accumulation in mouse cerebellar slices. Brain Res. 1985;343(1):129–36.
75. Errico F, Pirro MT, Affuso A, Spinelli P, De Felice M, D'Aniello A, et al. A physiological mechanism to regulate D-aspartic acid and NMDA levels in mammals revealed by D-aspartate oxidase deficient mice. Gene. 2006;374:50–7.
76. Errico F, Nistico R, Di Giorgio A, Squillace M, Vitucci D, Galbusera A, et al. Free D-aspartate regulates neuronal dendritic morphology, synaptic plasticity, gray matter volume and brain activity in mammals. Transl Psychiatry. 2014;4:e417.
77. Moghaddam B, Javitt D. From revolution to evolution: the glutamate hypothesis of schizophrenia and its implication for treatment. Neuropsychopharmacology. 2012;37(1):4–15.
78. Coyle JT. NMDA receptor and schizophrenia: a brief history. Schizophr Bull. 2012;38(5):920–6.
79. Coyle JT, Tsai G. NMDA receptor function, neuroplasticity, and the pathophysiology of schizophrenia. Int Rev Neurobiol. 2004;59:491–515.
80. Hashimoto K. Targeting of NMDA receptors in new treatments for schizophrenia. Expert Opin Ther Targets. 2014;18(9):1049–63.
81. Hashimoto K, Malchow B, Falkai P, Schmitt A. Glutamate modulators as potential therapeutic drugs in schizophrenia and affective disorders. Eur Arch Psychiatry Clin Neurosci. 2013;263(5):367–77.
82. Balu DT, Coyle JT. The NMDA receptor 'glycine modulatory site' in schizophrenia: D-serine, glycine, and beyond. Curr Opin Pharmacol. 2015;20:109–15.
83. Gottesman II, Gould TD. The endophenotype concept in psychiatry: etymology and strategic intentions. Am J Psychiatry. 2003;160(4):636–45.
84. Geyer MA. The family of sensorimotor gating disorders: comorbidities or diagnostic overlaps? Neurotox Res. 2006;10(3–4):211–20.
85. Errico F, D'Argenio V, Sforazzini F, Iasevoli F, Squillace M, Guerri G, et al. A role for D-aspartate oxidase in schizophrenia and in schizophrenia-related symptoms induced by phencyclidine in mice. Transl Psychiatry. 2015;5:e512.
86. Meyer-Lindenberg A. From maps to mechanisms through neuroimaging of schizophrenia. Nature. 2010;468(7321):194–202.
87. Stephan KE, Baldeweg T, Friston KJ. Synaptic plasticity and dysconnection in schizophrenia. Biol Psychiatry. 2006;59(10):929–39.
88. Zhou Y, Shu N, Liu Y, Song M, Hao Y, Liu H, et al. Altered resting-state functional connectivity and anatomical connectivity of hippocampus in schizophrenia. Schizophr Res. 2008;100(1–3):120–32.
89. Centonze D, Usiello A, Costa C, Picconi B, Erbs E, Bernardi G, et al. Chronic haloperidol promotes corticostriatal long-term potentiation by targeting dopamine D2L receptors. J Neurosci. 2004;24(38):8214–22.
90. Arnt J, Skarsfeldt T. Do novel antipsychotics have similar pharmacological characteristics? A review of the evidence. Neuropsychopharmacology. 1998;18(2):63–101.
91. Errico F, Napolitano F, Squillace M, Vitucci D, Blasi G, de Bartolomeis A, et al. Decreased levels of d-aspartate and NMDA in the prefrontal cortex and striatum of patients with schizophrenia. J Psychiatr Res. 2013;47(10):1432–7.
92. Colantuoni C, Lipska BK, Ye T, Hyde TM, Tao R, Leek JT, et al. Temporal dynamics and genetic control of transcription in the human prefrontal cortex. Nature. 2011;478(7370):519–23.
93. Hardingham GE, Bading H. The Yin and Yang of NMDA receptor signalling. Trends Neurosci. 2003;26(2):81–9.
94. Lancelot E, Beal MF. Glutamate toxicity in chronic neurodegenerative disease. Prog Brain Res. 1998;116:331–47.

95. Cristino L, Luongo L, Squillace M, Paolone G, Mango D, Piccinin S, et al. d-Aspartate oxidase influences glutamatergic system homeostasis in mammalian brain. Neurobiol Aging. 2015;36(5):1890–902.
96. Fatemi SH, Folsom TD. The neurodevelopmental hypothesis of schizophrenia, revisited. Schizophr Bull. 2009;35(3):528–48.
97. Owen MJ, O'Donovan MC, Thapar A, Craddock N. Neurodevelopmental hypothesis of schizophrenia. Br J Psychiatry. 2011;198(3):173–5.
98. Lu L, Mamiya T, Koseki T, Mouri A, Nabeshima T. Genetic animal models of schizophrenia related with the hypothesis of abnormal neurodevelopment. Biol Pharm Bull. 2011;34(9):1358–63.
99. Ikonomidou C, Bittigau P, Koch C, Genz K, Hoerster F, Felderhoff-Mueser U, et al. Neurotransmitters and apoptosis in the developing brain. Biochem Pharmacol. 2001;62(4):401–5.
100. Nacher J, McEwen BS. The role of N-methyl-D-asparate receptors in neurogenesis. Hippocampus. 2006;16(3):267–70.
101. Ritter LM, Vazquez DM, Meador-Woodruff JH. Ontogeny of ionotropic glutamate receptor subunit expression in the rat hippocampus. Brain Res Dev Brain Res. 2002;139(2):227–36.

Chapter 6
NMDA Receptors and Signaling in Chronic Neuropathic Pain

Geoffroy Laumet, Shao-Rui Chen, and Hui-Lin Pan

Abstract Chronic pain, especially neuropathic pain resulting from damage to the peripheral or central nervous system, is a major clinical problem that remains difficult to treat. The glutamate *N*-methyl-D-aspartate receptors (NMDARs) are widely distributed along the pain pathway and are critically involved in synaptic plasticity in chronic pain states. Although NMDAR antagonists, such as ketamine and memantine, have little effect on normal nociception, they are effective in treating neuropathic pain in animal models and in humans. Increased presynaptic NMDAR activity at primary afferent terminals potentiates excitatory input to spinal dorsal horn neurons, while increased postsynaptic NMDAR activity can increase neuronal excitability and diminish synaptic inhibition through promoting K^+-Cl^- cotransporter-2 proteolysis. Identifying the molecular mechanisms behind this increase in NMDAR activity, including protein phosphorylation and protein-protein interactions, could facilitate the development of new drugs that specifically reverse abnormal NMDAR hyperactivity in chronic pain with minimal impairment of the physiological functions of NMDARs.

Keywords Ion channels • NMDA receptors • Synaptic transmission • Synaptic plasticity • Pain • Neuropathic pain • Spinal cord • Ketamine

G. Laumet • S.-R. Chen
Department of Anesthesiology and Perioperative Medicine, Center for Neuroscience and Pain Research, The University of Texas MD Anderson Cancer Center, Houston, TX 77030, USA

H.-L. Pan (✉)
Department of Anesthesiology and Perioperative Medicine, Center for Neuroscience and Pain Research, The University of Texas MD Anderson Cancer Center, Houston, TX 77030, USA

Department of Anesthesiology and Perioperative Medicine, Unit 110, The University of Texas MD Anderson Cancer Center, 1515 Holcombe Blvd., Houston, TX 77030, USA
e-mail: huilinpan@mdanderson.org

K. Hashimoto (ed.), *The NMDA Receptors*, The Receptors 30,
DOI 10.1007/978-3-319-49795-2_6

Abbreviations

AMPAR	α-Amino-3-hydroxy-5-methyl-4-isoxazolepropionic acid receptor
AP5	(2R)-Amino-5-phosphonopentanoate
CaMKII	Ca2+/calmodulin-dependent kinase II
CIPS	Calcineurin inhibitor-induced pain syndrome
CK2	Casein kinase II
DRG	Dorsal root ganglion
GABA	γ-Aminobutyric acid
KCC2	K^+-Cl^- cotransporter-2
NMDAR	*N*-methyl-D-aspartate receptor
PKA	Protein kinase A
PKC	Protein kinase C

6.1 Introduction

In the mammalian central nervous system, excitatory synaptic transmission is mediated primarily by the amino acid glutamate. Fast synaptic transmission is principally mediated by α-amino-3-hydroxy-5-methyl-4-isoxazolepropionic acid receptors (AMPARs), whereas the more slowly activated *N*-methyl-D-aspartate receptors (NMDARs) mainly mediate synaptic plasticity. NMDARs are constitutively blocked by Mg^{2+} at resting membrane potentials. An interesting property of NMDARs is that their channel opening requires both glutamate binding and membrane depolarization, which is necessary to remove the Mg^{2+} block [1, 2]. Glycine was initially identified as the co-agonist of NMDARs [3]. However, it has since been shown that increased glycine release does not potentiate NMDAR activity in spinal dorsal horn neurons [4]. Notably, D-serine is a similarly potent NMDAR co-agonist that is required for postsynaptic NMDAR activity [5]; endogenous D-serine may be secreted mostly by astrocytes [6]. It has been shown that while glycine and D-serine are both endogenous co-agonists for NMDARs, they act at distinct populations of receptors, with D-serine present at synaptic NMDARs and glycine at extrasynaptic NMDARs [7].

NMDARs have three major subunits: GluN1, GluN2A-D (GluN2A, GluN2B, GluN2C and GluN2D), and GluN3A-B (GluN3A and GluN3B) [8, 9]. The core NMDARs contain two obligate GluN1 subunits and two GluN2(A-D) or GluN3(A-B) subunits. The subunit composition determines the pharmacological and physiological properties of the NMDAR. For example, NMDARs containing GluN2C and GluN2D are less sensitive to Mg^{2+} block than are NMDARs containing GluN2A or GluN2B [9, 10]. In addition, the tetrameric channel containing GluN1 and GluN2A displays faster inactivation than do channels composed of GluN1 and GluN2B, GluN2C, or GluN2D. Finally, GluN2A-D bind to glutamate, whereas GluN1 and

GluN3 bind to glycine. The NMDAR channel is embedded in a multiprotein complex. Proteins associated with the core NMDARs have important roles in the trafficking, stability, subunit composition, and function of NMDARs and are also involved in synaptic plasticity [1, 11, 12]. Although NMDARs do not play an important role in normal nociception, they are critically involved in the synaptic plasticity associated with the development of chronic pain. In this chapter, we review the current evidence about the changes in synaptic NMDAR activity that take place in neuropathic pain and the clinical utility of NMDAR antagonists for treatment of several chronic pain conditions.

6.2 NMDAR Distribution Along the Pain Pathway

The transmission of nociceptive signals from the spinal dorsal horn to the brain is referred to as the ascending pathway (Fig. 6.1.). Noxious stimuli are detected by the nerve endings of primary sensory neurons (nociceptors). The sensory nerves are endowed with specific receptors and channels that recognize noxious thermal, chemical, or mechanical stimuli. Nociceptive signals at the peripheral nerve endings are conveyed along axons in the form of action potentials. The cell bodies of primary sensory neurons are located in the dorsal root ganglion (DRG) and trigeminal ganglion. The central terminals of the sensory neurons end in in the spinal dorsal horn, where they release excitatory neurotransmitters, such as glutamate and substance P [13] (Fig. 6.1.).

Presynaptic NMDARs are expressed in DRG neurons and their central terminals. Immunocytochemistry studies have identified GluN1 in primary afferent nerves [14, 15] and found that more than 70 % of GluN1-labeled terminals are also positive for glutamate [14]. GluN2B, GluN2C, and GluN2D, but not GluN2A, have also been found at the central terminals of DRG neurons [16]. RNA sequencing data have shown the presence of mRNAs encoding GluN1 and GluN2B, but only very low mRNA levels of GluN2A, GluN2C, and GluN2D, in rat DRGs [17]. Interestingly, these RNA sequencing data also revealed a high level of GluN3A and GluN3B mRNAs in the DRG [17], but the physiological role of GluN3 subunits in regulating nociception remains unclear.

In the spinal dorsal horn, postsynaptic NMDARs are arranged at the postsynaptic density by a scaffold of the membrane-associated protein PSD-95 (postsynaptic density of 95 kDa). PSD-95 binds directly to NMDARs [18]. In the spinal cord, most NMDARs contain two GluN1 subunits, two GluN2 subunits, and, in some cases, a GluN3 subunit [19, 20]. GluN1, GluN2A, GluN2B, and GluN2D are all present in the superficial dorsal horn. GluN1 is expressed in all laminae of the spinal cord [21], and GluN2B is mostly distributed in the superficial dorsal horn. While GluN2A is predominantly concentrated at the synaptic level, GluN2B is also located at extrasynaptic sites [22]. The GluN1 subunit is encoded by only one gene, *Grin1*, but numerous variants of *Grin1* are produced by alternative splicing in the dorsal spinal cord [23].

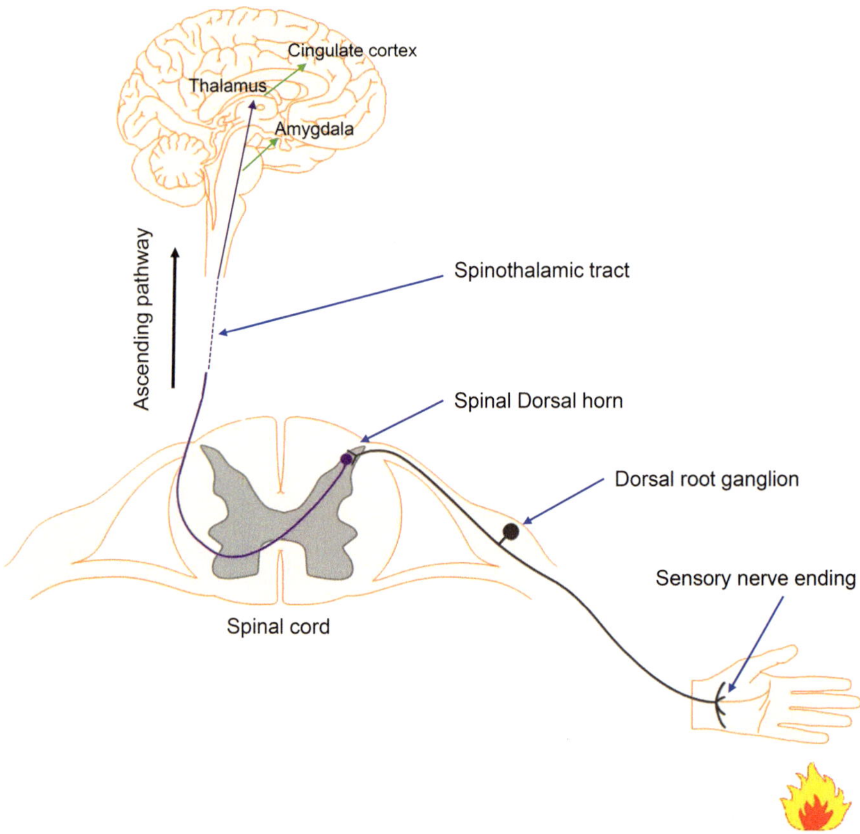

Fig. 6.1 Nociceptive transmission from nociceptors to the spinal cord and brain. A noxious stimulus initially activates the nerve endings of the primary sensory neurons, which convey the information to the second-order neurons in the spinal dorsal horn. The dorsal horn neurons then transmit the nociceptive information through the ascending pathways (i.e., the spinothalamic tract) to the brain. Several brain regions, including the amygdala and cingulate cortex, are involved in processing and eliciting pain perception

The conscious sensation of pain, a complex process influencing mood, cognition, emotion, and social behavior, involves many brain regions [24, 25]. Excitatory glutamatergic transmission in the cingulate cortex and amygdala is involved in the affective component of pain [26, 27] and pain-related negative emotions [28]. Various NMDAR subunits are distributed throughout these brain regions, but their contribution to the affective and emotional aspects of pain is uncertain.

6.3 Role of NMDARs in Chronic Pain in Animal Models

The activity-dependent synaptic plasticity in the spinal dorsal horn that is responsible for peripheral nerve injury-induced pain hypersensitivity is termed *central sensitization*. In neuropathic pain, the induction and maintenance of central sensitization is largely dependent on NMDAR activity [29]. Studies of various chronic pain models have clearly demonstrated that NMDARs at the spinal cord level play a critical role in the development of pain hypersensitivity. Early studies reported that intrathecal injection of NMDAR channel blockers, such as MK-801, ketamine, memantine, and dextromethorphan, had no effect on the nociceptive threshold in control animals but alleviated thermal hyperalgesia and mechanical allodynia in rats with peripheral nerve injury [30–33]. Another study found that intrathecal, but not intracerebral, injection of NMDAR antagonists diminished pain hypersensitivity induced by nerve injury, suggesting that neuropathic pain is mostly mediated by NMDAR hyperactivity at the spinal cord level [33]. These findings have subsequently been validated using animal models of several causes of neuropathic pain, including chronic constriction injury of the sciatic nerve [34, 35], spinal nerve ligation [36, 37], and peripheral neuropathy induced by treatment with the chemotherapeutic agent oxaliplatin [38]. The NMDAR channel blockers also profoundly reduced the evoked responses of spinal dorsal horn neurons in nerve-injured rats [39], confirming the central role of NMDAR activity in the spinal cord in generating chronic neuropathic pain.

Intrathecal administration of another NMDAR antagonist, (2R)-amino-5-phosphonopentanoate (AP5), also reversed the hypersensitivity to mechanically induced pain caused by spinal nerve injury in rats [40]. In an animal model of calcineurin inhibitor-induced pain syndrome (CIPS), AP5 attenuated, in a dose-dependent manner, tactile allodynia and mechanical hyperalgesia induced by tacrolimus (FK-506) [41]. Furthermore, intrathecal injection of AP5 significantly attenuated the development of analgesic tolerance and hyperalgesia caused by long-term morphine administration [42].

Systemic administration of memantine also reduced pain hypersensitivity in rat models of painful mononeuropathy [43] and painful diabetic neuropathy [44]. In a model of oxaliplatin-induced painful neuropathy, the effect of memantine seemed to be mediated mainly by spinal GluN2B-containing NMDARs [38]. Systemic administration of memantine also reversed mechanical pain hypersensitivity in a rat model of CIPS [41].

The limbic system, including the cingulate cortex, insula, and amygdala, is involved in the processing of the emotional states associated with chronic pain [45, 46]. Glutamatergic transmission in the cingulate cortex may be involved in affective pain [47]. Impaired motivation in mice with chronic pain appears to be mediated by GluN2B-containing NMDARs in the nucleus accumbens [48]. Chronic pain and depression may be neurochemically linked by the kynurenine pathway, which, through its byproduct quinolinic acid, an NMDAR agonist, is known to play a role in the development of inflammation-induced depression [49] and can be reduced by ketamine [50].

6.4 Clinical Use of NMDAR Antagonists in Treating Patients with Neuropathic Pain

NMDAR antagonists are generally effective for treating patients with certain neuropathic pain conditions, but chronic neuropathic pain remains a therapeutic challenge partly because neuropathic pain conditions have distinct etiologies and different underlying mechanisms [51]. Optimal treatment regimens should be determined based on the particular cause of neuropathic pain in the individual patient and the neurobiological mechanisms involved. The low-affinity NMDAR channel blockers, including ketamine, memantine, and dextromethorphan, are better tolerated in patients than is the high-affinity channel blocker MK-801 [52].

High-dose oral dextromethorphan is effective in the treatment of painful diabetic neuropathy, but not postherpetic neuralgia [53, 54]. Ketamine has been used to treat various chronic pain syndromes. Intravenous ketamine infusion is effective for the treatment of complex regional pain syndrome, especially for those patients for whom conventional analgesics have failed [55–58]. Infusion of low doses of ketamine also alleviates the severe pain associated with sickle cell disease [59, 60]. Common adverse effects of ketamine include hallucinations, memory defects, panic attacks, nausea/vomiting, and somnolence [61]. Nevertheless, ketamine seems to be tolerated by most patients [62]. Combining ketamine with other analgesics is a promising therapeutic strategy. A multi-day low-dose ketamine infusion as an adjuvant to oral gabapentin safely and efficaciously reduces spinal cord injury-induced chronic pain [63]. A combination of intravenous ketamine with opioids also improves pain management in patients with cancer [64].

Memantine, an orally active NMDAR antagonist, generally has fewer adverse effects because of its fast kinetics and relatively strong voltage dependence at the NMDAR channels. In one study, 8 weeks of memantine treatment reduced neuropathic pain induced by traumatic injury to the arm [65]. However, memantine seems to have little effect in patients with painful diabetic neuropathy and postherpetic neuralgia [54]. Placebo-controlled studies indicate that memantine is also ineffective for the treatment of chronic phantom limb pain and HIV-associated neuropathy [66, 67]. In a clinical study of traumatic amputation-induced neuropathic pain, memantine reduced pain prevalence at 6, but not 12, months after surgery [68]. However, in another clinical trial, 20 mg/day memantine had little effect on spontaneous and evoked pain hypersensitivity in patients with surgery-induced neuropathic pain [69]. A recent clinical trial showed that memantine effectively relieved pain and improved its emotional status in patients with fibromyalgia at 3- and 6-month follow-up and that the treatment was well tolerated [70]. Overall, it seems that memantine is no more efficacious than ketamine for treating neuropathic pain.

6.5 Changes in Synaptic NMDAR Activity in Neuropathic Pain

Presynaptic NMDARs are more resistant to Mg^{2+} blockage and action potential suppression than are postsynaptic NMDARs [71–74]. NMDAR activity can be recorded in DRG neurons [75, 76], but under physiological conditions, presynaptic NMDARs in the spinal dorsal horn are latent and not functionally active in regulating glutamate release from the central terminals of DRG neurons [42, 77]. Spinal nerve ligation in rats reduces the mRNA expression level of GluN1 but increases that of GluN2B in the DRG [17]. Increased GluN2B protein levels in the DRG have also been reported after nerve injury [78] and chronic opioid use [75].

Changes in presynaptic NMDAR activity can be measured by recording miniature excitatory postsynaptic currents and NMDAR-mediated monosynaptic excitatory postsynaptic currents evoked directly from stimulation of the dorsal root in spinal cord slices. Increased presynaptic NMDAR activity in chronic pain conditions can increase the release of excitatory neurotransmitters, including glutamate, from primary afferent terminals to the spinal dorsal horn neurons [77–79]. Acute or chronic opioid exposure also increases presynaptic NMDAR activity and facilitates the release of glutamate, which is critically involved in opioid-induced hyperalgesia, from primary sensory neurons [42, 80]. It seems that the observed nerve injury-induced increase in presynaptic NMDAR activity in the DRG is mainly mediated by increased GluN2B expression [17, 78], although GluN2A also mediates increased presynaptic NMDAR activity in animal models of opioid tolerance [42] and CIPS [41]. At the present time, the functional significance and mechanism of potentiated presynaptic NMDARs in neuropathic pain are not fully understood.

Changes in postsynaptic NMDAR activity in neuropathic pain have been assessed by puff application of NMDA directly to recorded spinal dorsal horn neurons. Spinal nerve ligation and CIPS both induce a large increase in postsynaptic NMDAR activity in the spinal lamina II neurons [41, 77, 79]. The increased postsynaptic NMDAR activity seems to be mediated primarily by GluN2A in the spinal nerve injury model [77] and by both GluN2A and GluN2B in the CIPS model [41]. Interestingly, the postsynaptic NMDAR activity of lamina II neurons is not significantly altered in rat models of diabetic neuropathy and postherpetic neuralgia [77]. In addition, direct spinal cord slice recordings show that paclitaxel treatment does not significantly alter the postsynaptic NMDAR activity of spinal dorsal horn neurons [81].

6.6 Molecular Mechanisms of Increased NMDAR Activity in Neuropathic Pain

In neuropathic pain, enhanced NMDAR activity is largely mediated by post-translational modification, such as phosphorylation and synaptic inhibition, of NMDAR subunits and the proteins that interact with them. Phosphorylation is a plastic mechanism for regulating the function and activity of proteins: the addition of a phosphate group by a protein kinase turns receptors and channels on, and the removal of a phosphate group by a protein phosphatase turns them off. NMDARs can be phosphorylated by protein kinase A (PKA), protein kinase C (PKC) [82], casein kinase II (CK2) [83], Ca^+ calmodulin-dependent protein kinase II (CaMKII) [84], and the tyrosine kinases Src [85] and Fyn [86]. The subunit GluN2B is the most tyrosine-phosphorylated protein in the nervous tissue; Tyr-1472 is its major phosphorylation site. Genetically modified knock-in mice in which Tyr-1472 is mutated to phenylalanine (Tyr-1472-Phe) show attenuated pain hypersensitivity caused by viral infection or nerve injury [87, 88], highlighting the essential role of NMDAR phosphorylation in the development of chronic pain. In this section, we review the roles of the major protein kinases and phosphatases (e.g., calcineurin) in chronic pain (Fig. 6.2.).

Calcineurin and CK2, a phosphatase and a kinase, respectively, appear to have opposite effects on NMDAR activity. Calcineurin is a Ca^{2+}- and calmodulin-dependent serine/threonine protein phosphatase expressed in the DRG [89] and spinal cord neurons [90]. Calcineurin may modulate NMDARs by dephosphorylation of GluN2A [91]. Calcineurin also shortens the opening time of NMDARs [92]. Inhibition of calcineurin induces long-lasting increases in presynaptic and postsynaptic NMDAR activity of spinal lamina II neurons and, therefore, in pain hypersensitivity [41].

In the hippocampus, CK2, a serine/threonine kinase, is activated in an NMDAR-dependent manner and contributes to enhanced synaptic NMDAR activity [93, 94]. In animal models of chronic pain, CK2 has been shown to be critically involved in increased NMDAR activity in the spinal dorsal horn [77, 95]. CK2α and CK2β subunits are upregulated in the spinal cord after nerve injury, and CK2 inhibitors completely normalize the postsynaptic NMDAR activity of spinal dorsal horn neurons after spinal nerve injury [77]. Pharmacological inhibition of CK2 and CK2β knockdown by short interfering RNA at the spinal cord level both significantly reverse mechanical pain hypersensitivity induced by spinal nerve ligation [77]. In a rat model of CIPS, intrathecal injection of CK2 inhibitors can normalize the presynaptic and postsynaptic NMDAR activity of spinal dorsal horn neurons that has been increased by a calcineurin inhibitor [95].

PKA- and PKC-mediated NMDAR phosphorylation also facilitates the generation of pain hypersensitivity. An increase in NMDAR phosphorylation in the spinal cord has been reported at both a PKC-dependent site, Ser-896 [96], and a PKA-dependent site, Ser-897 [97]. Phosphorylation by PKC reduces Mg^{2+} blockage of NMDAR by decreasing the receptor's affinity for extracellular Mg^{2+} [98]. However, PKC inhibition does not contribute to the enhanced postsynaptic NMDAR activity of spinal dorsal horn neurons that is induced by spinal nerve ligation [77]. Nevertheless, PKC is critically involved in the increased presynaptic NMDAR activity at primary afferent terminals in the spinal cord and, therefore, in the development of opioid-induced hyperalgesia and analgesic tolerance [42].

The Src tyrosine kinase family is composed of Src, Fyn, Yes, Lck, and Lyn, which are all widely expressed in the central nervous system [99] and regulate ion channel activity, including that of NMDARs [100]. Phosphorylation of GluN2A by

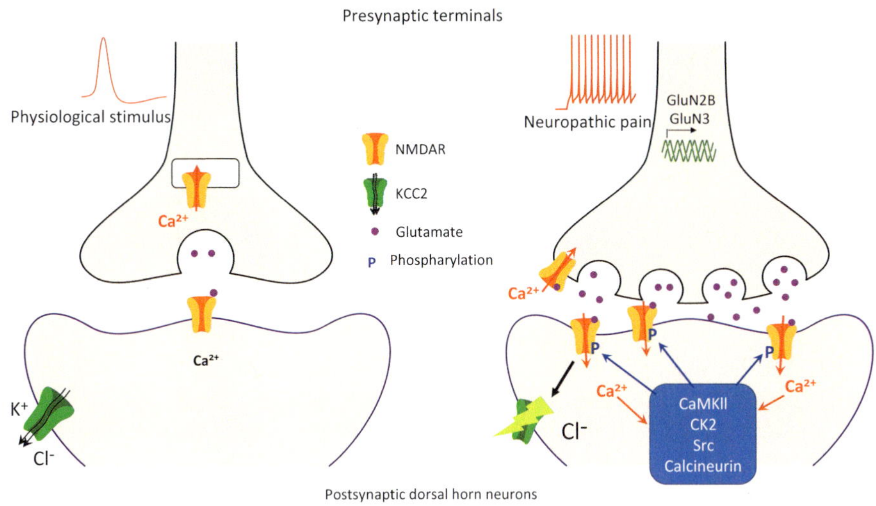

Fig. 6.2 Signaling mechanisms involved in regulating synaptic NMDAR activity in neuropathic pain. In neuropathic pain, increased presynaptic NMDAR activity promotes glutamate release from primary afferent nerve terminals and increases neuronal excitability and proteolytic cleavage of KCC2, which impairs synaptic inhibition. Protein kinases (e.g., CK2, CaMKII, and Src) and phosphatases (e.g., calcineurin) control NMDAR activity by regulating the phosphorylation of NMDARs and the proteins that interact with them

Src and Fyn kinases can potentiate NMDAR activity [12]. In the hippocampus, Tyr-1292, Tyr-Y1325, and Tyr-Y1387 of GluN2A are the major Src-mediated phosphorylation sites [101]. GluN2B is phosphorylated by tyrosine in the postsynaptic density [102]. Specific tyrosine residues in the C-terminal cytoplasmic region of GluN2B may be phosphorylated by Fyn [103]. Fyn-mediated GluN2B phosphorylation at Tyr-1472 is required for pain hypersensitivity development after peripheral nerve injury [86]. In the spinal cord, Src kinases are activated by peripheral nerve injury; injection of non-selective Src inhibitors reverses mechanical allodynia and hyperalgesia [104]. Tyrosine phosphorylation of GluN2B in the spinal dorsal horn is increased in animal models of both persistent [105, 106] and neuropathic pain [86]. Inhibition of Src kinases prevents the enhancement of GluN2B tyrosine phosphorylation [105, 106] and delays the onset of hyperalgesia [107]. Also, uncoupling Src and NMDARs with a peptide consisting of nine amino acids of Src fused with the Tat protein attenuates pain hypersensitivity in inflammatory and neuropathic pain models [108]. However, some evidence suggests that Src may contribute to neuropathic pain mainly through mechanisms other than NMDAR activation. For example, inhibition of Src kinase activity in the spinal cord has little effect on postsynaptic NMDAR activity that has been increased by nerve injury [77].

CaMKII is a Ca^{2+}- and calmodulin-dependent serine/threonine protein kinase that is widely expressed in the central nervous system. CaMKII interacts directly with GluN1 and GluN2 and contributes to activity-dependent synaptic plasticity [84]. GluN2B-containing NMDARs have a higher affinity to CaMKII than do GluN2A-containing NMDARs [11]. Given the major role of CaMKII in regulating synaptic plasticity, it is not surprising that spinal CaMKII plays a significant role in neuropathic pain. For example, intrathecal injection of a CaMKII inhibitor, KN-93, attenuates the development of thermal hyperalgesia and tactile allodynia after nerve injury [109–111]. Administration of KN-93 also alleviates pain hypersensitivity caused by bone cancer [112] and oxaliplatin-induced peripheral neuropathy [113]. Moreover, CaMKII phosphorylation in the spinal cord seems to be increased in models of neuropathic pain [87, 110, 112, 113]. Immunoblotting experiments indicate that CaMKII phosphorylates GluN2B mostly at the Ser-1303 site [114] and may alter GluN2B trafficking [112]. At the present time, however, there is no direct evidence demonstrating the role of spinal CaMKII in regulating presynaptic and postsynaptic NMDAR activity of spinal dorsal horn neurons in neuropathic pain models.

In sum, the balance between phosphorylation and dephosphorylation, which is mediated by the kinases and phosphatases, respectively, is critical to maintaining NMDAR activity. Dysregulation of this balance in favor of phosphorylation is likely to be an essential mechanism for potentiating NMDAR activity and for the transition from acute physiological pain to chronic pathological pain. Other post-transcriptional modifications, such as palmitoylation and ubiquitination, along with altered protein-protein interactions in NMDAR complexes, may also play a role in the increased NMDAR activity observed in chronic pain states and warrant further studies.

Besides dysregulation of phosphorylation, another mechanism involved in chronic pain involves synaptic inhibition. Peripheral nerve injury shifts γ-aminobutyric acid (GABA)- and glycine-mediated synaptic inhibition to excitation in spinal dorsal horn neurons [40, 115]. Activation of $GABA_A$ and glycine receptors normally inhibit dorsal horn neurons by controlling Cl^- influx. The low intracellular Cl^- concentration maintained by K^+-Cl^- cotransporter-2 (KCC2) hyperpolarizes the spinal dorsal horn neurons [116, 117]. The connection between NMDAR-mediated excitatory input and the loss of synaptic inhibition is not well understood. It has been shown, however, that spinal nerve injury promotes proteolytic cleavage of KCC2 by the protease calpain through NMDAR activation [40]. This finding suggests that NMDAR activation contributes to central sensitization not only by increasing neuronal excitability but also by diminishing synaptic inhibition through calpain-mediated KCC2 proteolysis (Fig. 6.2.). Remarkably, a recent study shows that restoring KCC2 expression levels in the spinal cord normalizes both presynaptic and postsynaptic NMDAR activity that has been increased by spinal nerve injury [79]. This reciprocal interaction between increased NMDAR activity and diminished synaptic inhibition and KCC2 function in the spinal dorsal horn may constitute a critical signaling mechanism for maintaining central sensitization and chronic neuropathic pain.

6.7 Conclusions and Future Directions

The role of NMDARs in the development of neuropathic pain has been well established in preclinical studies. Normalizing NMDAR activity, therefore, has been an attractive strategy for the treatment of chronic pain. Although NMDAR antagonists effectively reverse pain hypersensitivity in animal models of neuropathic pain, these drugs have variable effects in patients with neuropathic pain, possibly due to the heterogeneity of the sources of chronic pain in humans. It should be noted that preclinical research is conducted mostly using models of peripheral nerve injury, but neuropathic pain has many other etiologies (e.g., metabolic disorders, viral infection, and drug toxicity) in patients. Thus, the contribution of NMDARs to neuropathic pain needs to be investigated specifically using animal models of pain induced by means other than peripheral nerve injury. Moreover, spontaneous ongoing pain is usually measured in clinical studies [118], but preclinical research focuses predominantly on evoked-pain hypersensitivity. These differences in pain assessment methods may account for the differences in the results of clinical and animal studies on the efficacy of NMDAR antagonists in neuropathic pain. The possibility of a placebo effect should also be taken into account when interpreting the clinical data.

The recent development of subunit-selective NMDAR modulators may lead to improved treatment of neuropathic pain. Increased presynaptic NMDAR activity at primary afferent terminals potentiates excitatory input to spinal dorsal horn neurons, whereas increased postsynaptic NMDAR activity of spinal dorsal horn neurons can increase neuronal excitability and diminish synaptic inhibition through K^+-Cl^- cotransporter-2 proteolysis. The ideal way to treat chronic pain would be to specifically target the upstream mechanisms responsible for enhanced NMDAR activity in chronic pain without altering the physiological function of NMDARs. For example, a synthetic peptide blocking the interaction between Src and NMDARs has been found to reduce neuropathic pain [108]. Also, inhibition of CK2 normalizes NMDAR currents and pain hypersensitivity caused by nerve injury but does not affect normal NMDAR activity and nociceptive thresholds in naïve rats [77]. Thus, targeting the specific protein kinases or proteins that interact with NMDARs holds great promise for relieving chronic pain conditions without disturbing the physiological function of NMDARs.

References

1. Mayer ML, Armstrong N. Structure and function of glutamate receptor ion channels. Annu Rev Physiol. 2004;66:161–81.
2. Moriyoshi K, Masu M, Ishii T, Shigemoto R, Mizuno N, Nakanishi S. Molecular cloning and characterization of the rat NMDA receptor. Nature. 1991;354:31–7.
3. Johnson JW, Ascher P. Glycine potentiates the NMDA response in cultured mouse brain neurons. Nature. 1987;325:529–31.
4. Zhou HY, Zhang HM, Chen SR, Pan HL. Increased C-fiber nociceptive input potentiates inhibitory glycinergic transmission in the spinal dorsal horn. J Pharmacol Exp Ther. 2008;324:1000–10.
5. Mothet JP, Parent AT, Wolosker H, Brady Jr RO, Linden DJ, Ferris CD, Rogawski MA, Snyder SH. D-serine is an endogenous ligand for the glycine site of the N-methyl-D-aspartate receptor. Proc Natl Acad Sci U S A. 2000;97:4926–31.
6. Wolosker H. NMDA receptor regulation by D-serine: new findings and perspectives. Mol Neurobiol. 2007;36:152–64.

7. Papouin T, Ladepeche L, Ruel J, Sacchi S, Labasque M, Hanini M, Groc L, Pollegioni L, Mothet JP, Oliet SH. Synaptic and extrasynaptic NMDA receptors are gated by different endogenous coagonists. Cell. 2012;150:633–46.
8. Dingledine R, Borges K, Bowie D, Traynelis SF. The glutamate receptor ion channels. Pharmacol Rev. 1999;51:7–61.
9. Traynelis SF, Wollmuth LP, McBain CJ, Menniti FS, Vance KM, Ogden KK, Hansen KB, Yuan H, Myers SJ, Dingledine R. Glutamate receptor ion channels: structure, regulation, and function. Pharmacol Rev. 2010;62:405–96.
10. Wollmuth LP, Kuner T, Sakmann B. Adjacent asparagines in the NR2-subunit of the NMDA receptor channel control the voltage-dependent block by extracellular Mg2+. J Physiol. 1998;506:13–32.
11. Barria A, Malinow R. NMDA receptor subunit composition controls synaptic plasticity by regulating binding to CaMKII. Neuron. 2005;48:289–301.
12. Kohr G, Seeburg PH. Subtype-specific regulation of recombinant NMDA receptor-channels by protein tyrosine kinases of the src family. J Physiol. 1996;492:445–52.
13. Basbaum AI, Bautista DM, Scherrer G, Julius D. Cellular and molecular mechanisms of pain. Cell. 2009;139:267–84.
14. Liu H, Wang H, Sheng M, Jan LY, Jan YN, Basbaum AI. Evidence for presynaptic N-methyl-D-aspartate autoreceptors in the spinal cord dorsal horn. Proc Natl Acad Sci U S A. 1994;91:8383–7.
15. Sato K, Kiyama H, Park HT, Tohyama M. AMPA, KA and NMDA receptors are expressed in the rat DRG neurones. Neuroreport. 1993;4:1263–5.
16. Marvizon JC, McRoberts JA, Ennes HS, Song B, Wang X, Jinton L, Corneliussen B, Mayer EA. Two N-methyl-D-aspartate receptors in rat dorsal root ganglia with different subunit composition and localization. J Comp Neurol. 2002;446:325–41.
17. Laumet G, Garriga J, Chen SR, Zhang Y, Li DP, Smith TM, Dong Y, Jelinek J, Cesaroni M, Issa JP, Pan HL. G9a is essential for epigenetic silencing of K(+) channel genes in acute-to-chronic pain transition. Nat Neurosci. 2015;18:1746–55.
18. Sheng M, Hoogenraad CC. The postsynaptic architecture of excitatory synapses: a more quantitative view. Annu Rev Biochem. 2007;76:823–47.
19. Wee KS, Zhang Y, Khanna S, Low CM. Immunolocalization of NMDA receptor subunit NR3B in selected structures in the rat forebrain, cerebellum, and lumbar spinal cord. J Comp Neurol. 2008;509:118–35.
20. Wong HK, Liu XB, Matos MF, Chan SF, Perez-Otano I, Boysen M, Cui J, Nakanishi N, Trimmer JS, Jones EG, Lipton SA, Sucher NJ. Temporal and regional expression of NMDA receptor subunit NR3A in the mammalian brain. J Comp Neurol. 2002;450:303–17.
21. Boyce S, Wyatt A, Webb JK, O'Donnell R, Mason G, Rigby M, Sirinathsinghji D, Hill RG, Rupniak NM. Selective NMDA NR2B antagonists induce antinociception without motor dysfunction: correlation with restricted localisation of NR2B subunit in dorsal horn. Neuropharmacology. 1999;38:611–23.
22. Momiyama A. Distinct synaptic and extrasynaptic NMDA receptors identified in dorsal horn neurones of the adult rat spinal cord. J Physiol. 2000;523:621–8.
23. Tolle TR, Berthele A, Laurie DJ, Seeburg PH, Zieglgansberger W. Cellular and subcellular distribution of NMDAR1 splice variant mRNA in the rat lumbar spinal cord. Eur J Neurosci. 1995;7:1235–44.
24. Baliki MN, Apkarian AV. Nociception, pain, negative moods, and behavior selection. Neuron. 2015;87:474–91.
25. Talbot JD, Marrett S, Evans AC, Meyer E, Bushnell MC, Duncan GH. Multiple representations of pain in human cerebral cortex. Science. 1991;251:1355–8.
26. Johansen JP, Fields HL, Manning BH. The affective component of pain in rodents: direct evidence for a contribution of the anterior cingulate cortex. Proc Natl Acad Sci U S A. 2001;98:8077–82.
27. Lee CM, Sylantyev S, Shyu BC. Short-term synaptic plasticity in layer II/III of the rat anterior cingulate cortex. Brain Res Bull. 2007;71:416–27.

28. Neugebauer V, Galhardo V, Maione S, Mackey SC. Forebrain pain mechanisms. Brain Res Rev. 2009;60:226–42.
29. Woolf CJ, Thompson SW. The induction and maintenance of central sensitization is dependent on N-methyl-D-aspartic acid receptor activation; implications for the treatment of post-injury pain hypersensitivity states. Pain. 1991;44:293–9.
30. Malmberg AB, Yaksh TL. Hyperalgesia mediated by spinal glutamate or substance P receptor blocked by spinal cyclooxygenase inhibition. Science. 1992;257:1276–9.
31. Yamamoto T, Yaksh TL. Spinal pharmacology of thermal hyperesthesia induced by constriction injury of sciatic nerve. Excitatory amino acid antagonists. Pain. 1992;49:121–8.
32. Yamamoto T, Yaksh TL. Effects of intrathecal strychnine and bicuculline on nerve compression-induced thermal hyperalgesia and selective antagonism by MK-801. Pain. 1993;54:79–84.
33. Chaplan SR, Malmberg AB, Yaksh TL. Efficacy of spinal NMDA receptor antagonism in formalin hyperalgesia and nerve injury evoked allodynia in the rat. J Pharmacol Exp Ther. 1997;280:829–38.
34. Davar G, Hama A, Deykin A, Vos B, Maciewicz R. MK-801 blocks the development of thermal hyperalgesia in a rat model of experimental painful neuropathy. Brain Res. 1991;553:327–30.
35. Mao J, Price DD, Mayer DJ, Lu J, Hayes RL. Intrathecal MK-801 and local nerve anesthesia synergistically reduce nociceptive behaviors in rats with experimental peripheral mononeuropathy. Brain Res. 1992;576:254–62.
36. Leem JW, Choi EJ, Park ES, Paik KS. N-methyl-D-aspartate (NMDA) and non-NMDA glutamate receptor antagonists differentially suppress dorsal horn neuron responses to mechanical stimuli in rats with peripheral nerve injury. Neurosci Lett. 1996;211:37–40.
37. Leem JW, Kim HK, Hulsebosch CE, Gwak YS. Ionotropic glutamate receptors contribute to maintained neuronal hyperexcitability following spinal cord injury in rats. Exp Neurol. 2010;224:321–4.
38. Mihara Y, Egashira N, Sada H, Kawashiri T, Ushio S, Yano T, Ikesue H, Oishi R. Involvement of spinal NR2B-containing NMDA receptors in oxaliplatin-induced mechanical allodynia in rats. Mol Pain. 2011;7:8.
39. Suzuki R, Matthews EA, Dickenson AH. Comparison of the effects of MK-801, ketamine and memantine on responses of spinal dorsal horn neurones in a rat model of mononeuropathy. Pain. 2001;91:101–9.
40. Zhou HY, Chen SR, Byun HS, Chen H, Li L, Han HD, Lopez-Berestein G, Sood AK, Pan HL. N-methyl-D-aspartate receptor- and calpain-mediated proteolytic cleavage of K+–Cl- cotransporter-2 impairs spinal chloride homeostasis in neuropathic pain. J Biol Chem. 2012;287:33853–64.
41. Chen SR, Hu YM, Chen H, Pan HL. Calcineurin inhibitor induces pain hypersensitivity by potentiating pre- and postsynaptic NMDA receptor activity in spinal cords. J Physiol. 2014;592:215–27.
42. Zhao YL, Chen SR, Chen H, Pan HL. Chronic opioid potentiates presynaptic but impairs postsynaptic N-methyl-D-aspartic acid receptor activity in spinal cords: implications for opioid hyperalgesia and tolerance. J Biol Chem. 2012;287:25073–85.
43. Eisenberg E, LaCross S, Strassman AM. The clinically tested N-methyl-D-aspartate receptor antagonist memantine blocks and reverses thermal hyperalgesia in a rat model of painful mononeuropathy. Neurosci Lett. 1995;187:17–20.
44. Chen SR, Samoriski G, Pan HL. Antinociceptive effects of chronic administration of uncompetitive NMDA receptor antagonists in a rat model of diabetic neuropathic pain. Neuropharmacology. 2009;57:121–6.
45. Navratilova E, Atcherley CW, Porreca F. Brain Circuits Encoding Reward from Pain Relief. Trends Neurosci. 2015;38:741–50.
46. Neugebauer V, Li W, Bird GC, Han JS. The amygdala and persistent pain. Neuroscientist. 2004;10:221–34.
47. Li XY, Ko HG, Chen T, Descalzi G, Koga K, Wang H, Kim SS, Shang Y, Kwak C, Park SW, Shim J, Lee K, Collingridge GL, Kaang BK, Zhuo M. Alleviating neuropathic pain hypersensitivity by inhibiting PKMzeta in the anterior cingulate cortex. Science. 2010;330:1400–4.

48. Schwartz N, Temkin P, Jurado S, Lim BK, Heifets BD, Polepalli JS, Malenka RC. Chronic pain. Decreased motivation during chronic pain requires long-term depression in the nucleus accumbens. Science. 2014;345:535–42.
49. O'Connor JC, Lawson MA, Andre C, Moreau M, Lestage J, Castanon N, Kelley KW, Dantzer R. Lipopolysaccharide-induced depressive-like behavior is mediated by indoleamine 2,3-dioxygenase activation in mice. Mol Psychiatry. 2009;14:511–22.
50. Walker AK, Budac DP, Bisulco S, Lee AW, Smith RA, Beenders B, Kelley KW, Dantzer R. NMDA receptor blockade by ketamine abrogates lipopolysaccharide-induced depressive-like behavior in C57BL/6 J mice. Neuropsychopharmacology. 2013;38:1609–16.
51. Zhou HY, Chen SR, Pan HL. Targeting N-methyl-D-aspartate receptors for treatment of neuropathic pain. Expert Rev Clin Pharmacol. 2011;4:379–88.
52. Kalia LV, Kalia SK, Salter MW. NMDA receptors in clinical neurology: excitatory times ahead. Lancet Neurol. 2008;7:742–55.
53. Nelson KA, Park KM, Robinovitz E, Tsigos C, Max MB. High-dose oral dextromethorphan versus placebo in painful diabetic neuropathy and postherpetic neuralgia. Neurology. 1997;48:1212–8.
54. Sang CN, Booher S, Gilron I, Parada S, Max MB. Dextromethorphan and memantine in painful diabetic neuropathy and postherpetic neuralgia: efficacy and dose-response trials. Anesthesiology. 2002;96:1053–61.
55. Correll GE, Maleki J, Gracely EJ, Muir JJ, Harbut RE. Subanesthetic ketamine infusion therapy: a retrospective analysis of a novel therapeutic approach to complex regional pain syndrome. Pain Med. 2004;5:263–75.
56. Kiefer RT, Rohr P, Ploppa A, Dieterich HJ, Grothusen J, Koffler S, Altemeyer KH, Unertl K, Schwartzman RJ. Efficacy of ketamine in anesthetic dosage for the treatment of refractory complex regional pain syndrome: an open-label phase II study. Pain Med. 2008;9:1173–201.
57. Schwartzman RJ, Alexander GM, Grothusen JR, Paylor T, Reichenberger E, Perreault M. Outpatient intravenous ketamine for the treatment of complex regional pain syndrome: a double-blind placebo controlled study. Pain. 2009;147:107–15.
58. Sigtermans MJ, van Hilten JJ, Bauer MC, Arbous MS, Marinus J, Sarton EY, Dahan A. Ketamine produces effective and long-term pain relief in patients with Complex Regional Pain Syndrome Type 1. Pain. 2009;145:304–11.
59. Neri CM, Pestieau SR, Darbari DS. Low-dose ketamine as a potential adjuvant therapy for painful vaso-occlusive crises in sickle cell disease. Paediatr Anaesth. 2013;23:684–9.
60. Uprety D, Baber A, Foy M. Ketamine infusion for sickle cell pain crisis refractory to opioids: a case report and review of literature. Ann Hematol. 2014;93:769–71.
61. Niesters M, Martini C, Dahan A. Ketamine for chronic pain: risks and benefits. Br J Clin Pharmacol. 2014;77:357–67.
62. Azari P, Lindsay DR, Briones D, Clarke C, Buchheit T, Pyati S. Efficacy and safety of ketamine in patients with complex regional pain syndrome: a systematic review. CNS Drugs. 2012;26:215–28.
63. Amr YM. Multi-day low dose ketamine infusion as adjuvant to oral gabapentin in spinal cord injury related chronic pain: a prospective, randomized, double blind trial. Pain Physician. 2010;13:245–9.
64. Mercadante S, Arcuri E, Tirelli W, Casuccio A. Analgesic effect of intravenous ketamine in cancer patients on morphine therapy: a randomized, controlled, double-blind, crossover, double-dose study. J Pain Symptom Manage. 2000;20:246–52.
65. Sinis N, Birbaumer N, Gustin S, Schwarz A, Bredanger S, Becker ST, Unertl K, Schaller HE, Haerle M. Memantine treatment of complex regional pain syndrome: a preliminary report of six cases. Clin J Pain. 2007;23:237–43.
66. Schifitto G, Yiannoutsos CT, Simpson DM, Marra CM, Singer EJ, Kolson DL, Nath A, Berger JR, Navia B. A placebo-controlled study of memantine for the treatment of human immunodeficiency virus-associated sensory neuropathy. J Neurovirol. 2006;12:328–31.

67. Wiech K, Kiefer RT, Topfner S, Preissl H, Braun C, Unertl K, Flor H, Birbaumer N. A placebo-controlled randomized crossover trial of the N-methyl-D-aspartic acid receptor antagonist, memantine, in patients with chronic phantom limb pain. Anesth Analg. 2004;98:408–13.
68. Schley M, Topfner S, Wiech K, Schaller HE, Konrad CJ, Schmelz M, Birbaumer N. Continuous brachial plexus blockade in combination with the NMDA receptor antagonist memantine prevents phantom pain in acute traumatic upper limb amputees. Eur J Pain. 2007;11:299–308.
69. Nikolajsen L, Gottrup H, Kristensen AG, Jensen TS. Memantine (a N-methyl-D-aspartate receptor antagonist) in the treatment of neuropathic pain after amputation or surgery: a randomized, double-blinded, cross-over study. Anesth Analg. 2000;91:960–6.
70. Olivan-Blazquez B, Herrera-Mercadal P, Puebla-Guedea M, Perez-Yus MC, Andres E, Fayed N, Lopez-Del-Hoyo Y, Magallon R, Roca M, Garcia-Campayo J. Efficacy of memantine in the treatment of fibromyalgia: a double-blind, randomised, controlled trial with 6-month follow-up. Pain. 2014;155:2517–25.
71. Bender VA, Bender KJ, Brasier DJ, Feldman DE. Two coincidence detectors for spike timing-dependent plasticity in somatosensory cortex. J Neurosci. 2006;26:4166–77.
72. Brasier DJ, Feldman DE. Synapse-specific expression of functional presynaptic NMDA receptors in rat somatosensory cortex. J Neurosci. 2008;28:2199–211.
73. Gill I, Droubi S, Giovedi S, Fedder KN, Bury LA, Bosco F, Sceniak MP, Benfenati F, Sabo SL. Presynaptic NMDA receptors - dynamics and distribution in developing axons in vitro and in vivo. J Cell Sci. 2015;128:768–80.
74. Woodhall G, Evans DI, Cunningham MO, Jones RS. NR2B-containing NMDA autoreceptors at synapses on entorhinal cortical neurons. J Neurophysiol. 2001;86:1644–51.
75. Gong K, Bhargava A, Jasmin L. GluN2B N-methyl-D-aspartate receptor and excitatory amino acid transporter 3 are upregulated in primary sensory neurons after 7 days of morphine administration in rats: implication for opiate-induced hyperalgesia. Pain. 2016;157:147–58.
76. Li J, McRoberts JA, Nie J, Ennes HS, Mayer EA. Electrophysiological characterization of N-methyl-D-aspartate receptors in rat dorsal root ganglia neurons. Pain. 2004;109:443–52.
77. Chen SR, Zhou HY, Byun HS, Chen H, Pan HL. Casein kinase II regulates N-methyl-D-aspartate receptor activity in spinal cords and pain hypersensitivity induced by nerve injury. J Pharmacol Exp Ther. 2014;350:301–12.
78. Yan X, Jiang E, Gao M, Weng HR. Endogenous activation of presynaptic NMDA receptors enhances glutamate release from the primary afferents in the spinal dorsal horn in a rat model of neuropathic pain. J Physiol. 2013;591:2001–19.
79. Li L, Chen SR, Chen H, Wen L, Hittelman WN, Pan HL. Chloride homeostasis critically regulates synaptic NMDA receptor activity in neuropathic pain. Cell Rep. 2016;15(7):1376–83.
80. Zhou HY, Chen SR, Chen H, Pan HL. Opioid-induced long-term potentiation in the spinal cord is a presynaptic event. J Neurosci. 2010;30:4460–6.
81. Chen SR, Zhu L, Chen H, Wen L, Laumet G, Pan HL. Increased spinal cord Na(+)-K(+)-2Cl(−) cotransporter-1 (NKCC1) activity contributes to impairment of synaptic inhibition in paclitaxel-induced neuropathic pain. J Biol Chem. 2014;289:31111–20.
82. Tingley WG, Ehlers MD, Kameyama K, Doherty C, Ptak JB, Riley CT, Huganir RL. Characterization of protein kinase A and protein kinase C phosphorylation of the N-methyl-D-aspartate receptor NR1 subunit using phosphorylation site-specific antibodies. J Biol Chem. 1997;272:5157–66.
83. Lieberman DN, Mody I. Casein kinase-II regulates NMDA channel function in hippocampal neurons. Nat Neurosci. 1999;2:125–32.
84. Leonard AS, Lim IA, Hemsworth DE, Horne MC, Hell JW. Calcium/calmodulin-dependent protein kinase II is associated with the N-methyl-D-aspartate receptor. Proc Natl Acad Sci U S A. 1999;96:3239–44.
85. Yu XM, Askalan R, Keil 2nd GJ, Salter MW. NMDA channel regulation by channel-associated protein tyrosine kinase Src. Science. 1997;275:674–8.

86. Abe T, Matsumura S, Katano T, Mabuchi T, Takagi K, Xu L, Yamamoto A, Hattori K, Yagi T, Watanabe M, Nakazawa T, Yamamoto T, Mishina M, Nakai Y, Ito S. Fyn kinase-mediated phosphorylation of NMDA receptor NR2B subunit at Tyr1472 is essential for maintenance of neuropathic pain. Eur J Neurosci. 2005;22:1445–54.
87. Matsumura S, Kunori S, Mabuchi T, Katano T, Nakazawa T, Abe T, Watanabe M, Yamamoto T, Okuda-Ashitaka E, Ito S. Impairment of CaMKII activation and attenuation of neuropathic pain in mice lacking NR2B phosphorylated at Tyr1472. Eur J Neurosci. 2010;32:798–810.
88. Unezaki S, Sasaki A, Mabuchi T, Matsumura S, Katano T, Nakazawa T, Nishio N, Andoh T, Yamamoto T, Nakatsuka T, Kuraishi Y, Ito S. Involvement of Tyr1472 phosphorylation of NMDA receptor NR2B subunit in postherpetic neuralgia in model mice. Mol Pain. 2012;8:59.
89. Wu ZZ, Chen SR, Pan HL. Transient receptor potential vanilloid type 1 activation down-regulates voltage-gated calcium channels through calcium-dependent calcineurin in sensory neurons. J Biol Chem. 2005;280:18142–51.
90. Strack S, Wadzinski BE, Ebner FF. Localization of the calcium/calmodulin-dependent protein phosphatase, calcineurin, in the hindbrain and spinal cord of the rat. J Comp Neurol. 1996;375:66–76.
91. Krupp JJ, Vissel B, Thomas CG, Heinemann SF, Westbrook GL. Calcineurin acts via the C-terminus of NR2A to modulate desensitization of NMDA receptors. Neuropharmacology. 2002;42:593–602.
92. Lieberman DN, Mody I. Regulation of NMDA channel function by endogenous Ca(2+)-dependent phosphatase. Nature. 1994;369:235–9.
93. Charriaut-Marlangue C, Otani S, Creuzet C, Ben-Ari Y, Loeb J. Rapid activation of hippocampal casein kinase II during long-term potentiation. Proc Natl Acad Sci U S A. 1991;88:10232–6.
94. Kimura R, Matsuki N. Protein kinase CK2 modulates synaptic plasticity by modification of synaptic NMDA receptors in the hippocampus. J Physiol. 2008;586:3195–206.
95. Hu YM, Chen SR, Chen H, Pan HL. Casein kinase II inhibition reverses pain hypersensitivity and potentiated spinal N-methyl-D-aspartate receptor activity caused by calcineurin inhibitor. J Pharmacol Exp Ther. 2014;349:239–47.
96. Brenner GJ, Ji RR, Shaffer S, Woolf CJ. Peripheral noxious stimulation induces phosphorylation of the NMDA receptor NR1 subunit at the PKC-dependent site, serine-896, in spinal cord dorsal horn neurons. Eur J Neurosci. 2004;20:375–84.
97. Zou X, Lin Q, Willis WD. Role of protein kinase A in phosphorylation of NMDA receptor 1 subunits in dorsal horn and spinothalamic tract neurons after intradermal injection of capsaicin in rats. Neuroscience. 2002;115:775–86.
98. Chen L, Huang LY. Protein kinase C reduces Mg2+ block of NMDA-receptor channels as a mechanism of modulation. Nature. 1992;356:521–3.
99. Cotton PC, Brugge JS. Neural tissues express high levels of the cellular src gene product pp60c-src. Mol Cell Biol. 1983;3:1157–62.
100. Wang YT, Salter MW. Regulation of NMDA receptors by tyrosine kinases and phosphatases. Nature. 1994;369:233–5.
101. Yang M, Leonard JP. Identification of mouse NMDA receptor subunit NR2A C-terminal tyrosine sites phosphorylated by coexpression with v-Src. J Neurochem. 2001;77:580–8.
102. Moon IS, Apperson ML, Kennedy MB. The major tyrosine-phosphorylated protein in the postsynaptic density fraction is N-methyl-D-aspartate receptor subunit 2B. Proc Natl Acad Sci U S A. 1994;91:3954–8.
103. Nakazawa T, Komai S, Tezuka T, Hisatsune C, Umemori H, Semba K, Mishina M, Manabe T, Yamamoto T. Characterization of Fyn-mediated tyrosine phosphorylation sites on GluR epsilon 2 (NR2B) subunit of the N-methyl-D-aspartate receptor. J Biol Chem. 2001;276:693–9.
104. Katsura H, Obata K, Mizushima T, Sakurai J, Kobayashi K, Yamanaka H, Dai Y, Fukuoka T, Sakagami M, Noguchi K. Activation of Src-family kinases in spinal microglia contributes to mechanical hypersensitivity after nerve injury. J Neurosci. 2006;26:8680–90.

105. Guo W, Wei F, Zou S, Robbins MT, Sugiyo S, Ikeda T, Tu JC, Worley PF, Dubner R, Ren K. Group I metabotropic glutamate receptor NMDA receptor coupling and signaling cascade mediate spinal dorsal horn NMDA receptor 2B tyrosine phosphorylation associated with inflammatory hyperalgesia. J Neurosci. 2004;24:9161–73.
106. Guo W, Zou S, Guan Y, Ikeda T, Tal M, Dubner R, Ren K. Tyrosine phosphorylation of the NR2B subunit of the NMDA receptor in the spinal cord during the development and maintenance of inflammatory hyperalgesia. J Neurosci. 2002;22:6208–17.
107. Sato E, Takano Y, Kuno Y, Takano M, Sato I. Involvement of spinal tyrosine kinase in inflammatory and N-methyl-D-aspartate-induced hyperalgesia in rats. Eur J Pharmacol. 2003;468:191–8.
108. Liu XJ, Gingrich JR, Vargas-Caballero M, Dong YN, Sengar A, Beggs S, Wang SH, Ding HK, Frankland PW, Salter MW. Treatment of inflammatory and neuropathic pain by uncoupling Src from the NMDA receptor complex. Nat Med. 2008;14:1325–32.
109. Dai Y, Wang H, Ogawa A, Yamanaka H, Obata K, Tokunaga A, Noguchi K. Ca2+/calmodulin-dependent protein kinase II in the spinal cord contributes to neuropathic pain in a rat model of mononeuropathy. Eur J Neurosci. 2005;21:2467–74.
110. Katano T, Nakazawa T, Nakatsuka T, Watanabe M, Yamamoto T, Ito S. Involvement of spinal phosphorylation cascade of Tyr1472-NR2B, Thr286-CaMKII, and Ser831-GluR1 in neuropathic pain. Neuropharmacology. 2011;60:609–16.
111. Chen Y, Luo F, Yang C, Kirkmire CM, Wang ZJ. Acute inhibition of Ca2+/calmodulin-dependent protein kinase II reverses experimental neuropathic pain in mice. J Pharmacol Exp Ther. 2009;330:650–9.
112. Liu Y, Liang Y, Hou B, Liu M, Yang X, Liu C, Zhang J, Zhang W, Ma Z, Gu X. The inhibitor of calcium/calmodulin-dependent protein kinase II KN93 attenuates bone cancer pain via inhibition of KIF17/NR2B trafficking in mice. Pharmacol, Biochem Behav. 2014;124:19–26.
113. Shirahama M, Ushio S, Egashira N, Yamamoto S, Sada H, Masuguchi K, Kawashiri T, Oishi R. Inhibition of Ca2+/calmodulin-dependent protein kinase II reverses oxaliplatin-induced mechanical allodynia in rats. Mol Pain. 2012;8:26.
114. Suo ZW, Fan QQ, Yang X, Hu XD. Ca2+/calmodulin-dependent protein kinase II in spinal dorsal horn contributes to the pain hypersensitivity induced by gamma-aminobutyric acid type a receptor inhibition. J Neurosci Res. 2013;91:1473–82.
115. Coull JA, Boudreau D, Bachand K, Prescott SA, Nault F, Sik A, De Koninck P, De Koninck Y. Trans-synaptic shift in anion gradient in spinal lamina I neurons as a mechanism of neuropathic pain. Nature. 2003;424:938–42.
116. Boulenguez P, Liabeuf S, Bos R, Bras H, Jean-Xavier C, Brocard C, Stil A, Darbon P, Cattaert D, Delpire E, Marsala M, Vinay L. Down-regulation of the potassium-chloride cotransporter KCC2 contributes to spasticity after spinal cord injury. Nat Med. 2010;16:302–7.
117. Hubner CA, Stein V, Hermans-Borgmeyer I, Meyer T, Ballanyi K, Jentsch TJ. Disruption of KCC2 reveals an essential role of K-Cl cotransport already in early synaptic inhibition. Neuron. 2001;30:515–24.
118. Treede RD, Jensen TS, Campbell JN, Cruccu G, Dostrovsky JO, Griffin JW, Hansson P, Hughes R, Nurmikko T, Serra J. Neuropathic pain: redefinition and a grading system for clinical and research purposes. Neurology. 2008;70:1630–5.

Chapter 7
Role of NMDA Receptors in Pancreatic Islets

Okka Scholz, Alena Welters, and Eckhard Lammert

Abstract Pancreatic islets are cell aggregates that consist of a few dozen to a few thousand endocrine pancreatic cells, primarily β-cells and α-cells. The latter secrete polypeptide hormones into the bloodstream, i.e. insulin and glucagon, to temporally decrease or increase blood glucose, respectively. Therefore, islets are essential for maintaining the concentration of glucose, the key nutrient of the central nervous system. Recently, *N*-methyl-D-aspartate receptors (NMDARs) were shown to reduce the amount of insulin secreted by β-cells in response to stimulatory glucose concentrations. Likewise, NMDAR deletion in mouse islets increases glucose-induced plasma insulin concentrations and lowers blood glucose. NMDAR expression in β-cells also promotes cell death under inflammatory and diabetogenic conditions. It is hypothesized that NMDARs on β-cells are permanently occupied by glutamate derived from the leaky, fenestrated blood capillary network of the islets, and that NMDAR activity is mainly induced by depolarization of the β-cells. The latter explains why deletion of NMDARs only increases insulin release from β-cells when blood glucose concentrations are high and β-cells are depolarized. Blocking NMDARs with the over-the-counter drug dextromethorphan (DXM) increases insulin release and lowers blood glucose concentrations in mice. In two clinical trials, DXM selectively increases serum insulin concentrations in individuals with type 2 diabetes mellitus and selectively lowers blood glucose whenever the blood glucose concentration is high. Therefore, NMDAR antagonists, and especially DXM, harbor anti-diabetic properties that increase β-cell function and survival.

O. Scholz (✉) • E. Lammert, Ph.D.
Institute of Metabolic Physiology, Heinrich Heine University Düsseldorf,
Universitätsstr. 1, 40225 Düsseldorf, Germany

Institute for Beta Cell Biology, German Diabetes Center, Leibniz Center for Diabetes Research Düsseldorf, Auf'm Hennekamp 65, 40225 Düsseldorf, Germany
e-mail: okka.scholz@hhu.de; lammert@hhu.de

A. Welters, M.D., Ph.D.
Institute of Metabolic Physiology, Heinrich Heine University Düsseldorf,
Universitätsstr. 1, 40225 Düsseldorf, Germany

Department of General Pediatrics, Neonatology and Pediatric Cardiology, University Children's Hospital Düsseldorf, Moorenstraße 5, 40225 Düsseldorf, Germany
e-mail: alena.welters@med.uni-duesseldorf.de

K. Hashimoto (ed.), *The NMDA Receptors*, The Receptors 30,
DOI 10.1007/978-3-319-49795-2_7

Keywords Blood glucose • Diabetes mellitus • Insulin • *N*-methyl-D-aspartate receptor • Pancreatic β-cell • Pancreatic islet

Abbreviations

AMPAR	α-amino-3-hydroxy-5-methyl-4-isoxazolepropionic acid receptor
BBB	Blood-brain barrier
CNS	Central nervous system
CREB	cAMP response element binding protein
DPP-4	Dipeptidyl peptidase-4
DXM	Dextromethorphan
DXO	Dextrorphan
GLUT	Glucose transporter
GSIS	Glucose-stimulated insulin secretion
NMDAR	*N*-methyl-D-aspartate receptor
NO	Nitric oxide
OGTT	Oral glucose tolerance test
SGLT-2	Sodium-glucose co-transporter-2
T1DM	Type 1 diabetes mellitus
T2DM	Type 2 diabetes mellitus
VDCC	Voltage-dependent Ca^{2+} channel

7.1 Introduction

The pancreas is a lobular gland with both exocrine and endocrine functionality [1]. The endocrine pancreas contains different hormone-secreting epithelial cell types that are clustered into small groups and scattered throughout the exocrine pancreas. These endocrine cell clusters are termed pancreatic islets, or islets of Langerhans. A human pancreas has around one million pancreatic islets [2], each consisting of 50–3000 endocrine cells [3], predominantly insulin-releasing β-cells (50 % of human pancreatic islets), glucagon-releasing α-cells (35–40 % of human pancreatic islets), and somatostatin-releasing δ-cells (10–15 % of human pancreatic islets) [2]. Hormones released by the endocrine pancreas, particularly insulin and glucagon, are essential to maintain blood glucose concentration within a narrow range, a process known as glucose homeostasis. Insulin is the only hormone that lowers blood glucose concentration [2]. In contrast, glucagon raises blood glucose concentration, but may also act on β-cells to stimulate insulin release [4]. Given this functionality, pancreatic islets are often referred to as 'mini-organs,' a claim further substantiated by their dense innervation and extensive vasculature [5, 6]. This connectivity facilitates intra-islet autocrine and paracrine interactions and, together with extra-pancreatic signals (e.g. from the gut, liver or adipose tissue), affects islet function and the release of islet cell hormones into the bloodstream [2, 6].

7.2 Pancreatic Beta Cells

7.2.1 Insulin Secretion

Blood glucose concentration increases following food ingestion and upon endogenous glucose production. High blood glucose concentrations trigger the release of insulin from pancreatic β-cells, a process termed glucose-stimulated insulin secretion (GSIS) [7, 8]. GSIS requires the intracellular uptake and metabolic degradation of glucose by pancreatic β-cells. Glucose enters the β-cell via low-affinity glucose transporters (GLUTs), specifically via GLUT1 and GLUT3 in humans [2]. Glucose metabolism in pancreatic β-cells results in an increase of the intracellular ATP/ADP ratio, which causes ATP-sensitive K^+ channels (K_{ATP} channels) to close and triggers plasma membrane depolarization, followed by an opening of voltage-dependent Ca^{2+} channels (VDCCs) [9]. Ca^{2+} influx via VDCCs increases intracellular Ca^{2+} in the form of Ca^{2+} oscillations, which trigger an oscillatory release of insulin [10, 11]. The latter is caused by fusion of insulin-containing secretory vesicles with the plasma membrane [2]. Ca^{2+} released from intracellular stores further increases intracellular Ca^{2+} concentrations and maintains insulin exocytosis [12]. Once released, the major target tissues of insulin are the liver, skeletal muscle, and adipose tissue [8]. In the latter two tissues, insulin promotes the translocation of GLUT4 from storage vesicles to the plasma membrane, whereby glucose uptake from the bloodstream takes place [13]. In the liver, insulin inhibits endogenous glucose production, promotes glycogen synthesis, and stimulates lipogenesis [7].

7.2.2 Glutamate as an Intra- and Extracellular Messenger

Pancreatic endocrine cells and central nervous system (CNS) neurons have many features in common, e.g. the expression of various receptors, signaling pathways, and cell adhesion molecules [14–16]. Glutamate signaling is crucial for normal brain function, as it is the primary excitatory neurotransmitter in the CNS [17]. Pancreatic β-cells express several components of the glutamate signaling system, e.g. vesicular glutamate transporters [18], excitatory amino acid transporters [19], and different cell surface receptors, including the ionotropic *N*-methyl-D-aspartate receptors (NMDARs) and α-amino-3-hydroxy-5-methyl-4-isoxazolepropionic acid receptors (AMPARs) [20–22], and metabotropic glutamate receptors [23]. Accumulating evidence suggests that both intracellular and extracellular glutamate are important for pancreatic islet physiology [24]. It has long been proposed that intracellular glutamate acts as a mitochondrial-derived messenger to enhance nutrient-stimulated insulin secretion in β-cells [25]. More recent data now indicates that the uptake of intracellular glutamate into insulin-containing secretory granules, which is accompanied by a lowering of the pH in these vesicles, underlies the stimulatory effects of incretins on insulin secretion [26]. Incretins are hormones produced by intestinal endocrine cells that bind to G-protein coupled receptors on β-cells in order to amplify GSIS [27]. Extracellular glutamate, which is available from the bloodstream, local α-cells, and possibly other pancreatic cell types [24], increases GSIS from islets via activation of AMPARs [28, 29]. It has for example been shown that glutamate is released together with glucagon from α-cells and activates AMPARs on the α-cells in an autocrine fashion to enhance glucagon release [30]. Importantly, glucagon also binds to glucagon receptors on β-cells and thereby increases GSIS [31, 32]. Since AMPARs are expressed on mouse β-cells, extracellular glutamate might also directly acts on the β-cells and activates insulin release [33].

Despite the numerous similarities, fundamental differences exist between glutamate signaling in the CNS and pancreatic islets. Most importantly, the blood-brain barrier (BBB) almost completely prevents free access of blood-derived glutamate to neurons [34]. This is important because increased extracellular glutamate concentrations are toxic to neurons [35]. In contrast, pancreatic islets are supplied by a fenestrated and highly permeable capillary network, which allows glutamate and most other plasma components to directly access β-cells [36]. Consequently, the β-cells are permanently exposed to high extracellular glutamate concentrations, and pancreatic NMDARs are expected to be fully saturated with glutamate [24]. Since opening of NMDARs requires two events, i.e. ligand binding (glutamate and co-agonist glycine/serine) and plasma membrane depolarization [37], and glutamate is permanently bound to NMDARs, β-cell depolarization (e.g. due to stimulatory high blood glucose concentrations) appears to be the critical event for NMDAR activation in β-cells.

Considering the numerous similarities between neurons and pancreatic endocrine cells, it is not surprising that pathologies affecting the CNS, i.e. neurodegenerative disorders, and diabetes mellitus share similar pathomechanisms, e.g. the formation of amyloid plaques [38], endoplasmatic reticulum stress [39] and the release of inflammatory cytokines leading to neuronal and islet cell death [40, 41]. In this context glutamate-mediated excitotoxicity is of particular interest. It is widely accepted that high extracellular glutamate concentrations and excessive NMDAR activity induce neuronal cell death, and that NMDAR dysfunction is associated with neurological disorders, including stroke and depression [42]. Likewise, high concentrations of extracellular glutamate have also been shown to be deleterious to the clonal β-cell line βTC3 and human pancreatic β-cells. This suggests that NMDAR-mediated excitotoxicity also plays a role in the pathogenesis of diabetes mellitus [18].

7.3 NMDARs in Beta Cells

7.3.1 *Expression of NMDARs in Beta Cells*

It has long been known that NMDARs are expressed in pancreatic islets and insulin-secreting β-cells [20–22], and application of NMDAR antagonists also suggested that NMDARs present on β-cells are functional, since insulin secretion and blood glucose control were affected [43–46]. The presence of functional NMDARs in pancreatic islets and β-cells has been confirmed by various experiments: for example, application of NMDA to pancreatic islets or insulinoma cells (a model for insulin-secreting β-cells) sometimes, but not always, results in altered insulin release [20–22]. However, the role of NMDARs in the pancreas remained elusive since studies investigating the *in vitro* or *in vivo* effects of either NMDAR activation or inhibition on insulin secretion and/or glucose tolerance had been contradictory [20–22, 45, 46].

Recent genetic studies have started to provide clarity. Pancreatic β-cell-specific deletion of the *Grin1* gene (encoding the obligatory GluN1 subunit of NMDARs) almost completely removes the GluN1 protein from mouse islets, indicating that the majority of islet NMDARs are expressed in β-cells. It was furthermore shown that NMDAR inhibition enhances GSIS from mouse and human pancreatic islets, improves glucose tolerance in mice and men, and promotes islet cell survival under inflammatory and diabetogenic conditions in both human pancreatic islets *in vitro* and in the type 2 diabetic mouse model db/db *in vivo* [40].

7.3.2 *Proposed Mechanism of NMDAR-Regulated Insulin Release*

By deleting *Grin1* from the pancreatic epithelium of mice, from which β-cells develop, a clear role for NMDARs in β-cell function emerged. The results demonstrate that NMDARs regulate GSIS and blood glucose homeostasis [40]. With this knowledge, the following model of NMDAR-mediated regulation of GSIS was proposed [24]: in islets perfused with blood, extracellular glutamate and co-agonist glycine (or serine) are bound to NMDARs. Under stimulatory high blood glucose concentrations, β-cell plasma membrane depolarization activates NMDARs (Fig. 7.1., step 1). As in some specific neurons, NMDAR activity promotes opening of K^+ channels, precisely Ca^{2+}-activated K^+ channels of intermediate conductance (SK4 channels) and K_{ATP} channels (Fig. 7.1., step 2) [47, 48]. Ca^{2+} influx through NMDARs may directly activate SK4 channels, likely within a small spatial domain (Ca^{2+} microdomain) [47] and may induce K_{ATP} channel opening via second messengers like nitric oxide (NO) [48, 49]. High concentrations of NO promote K_{ATP}

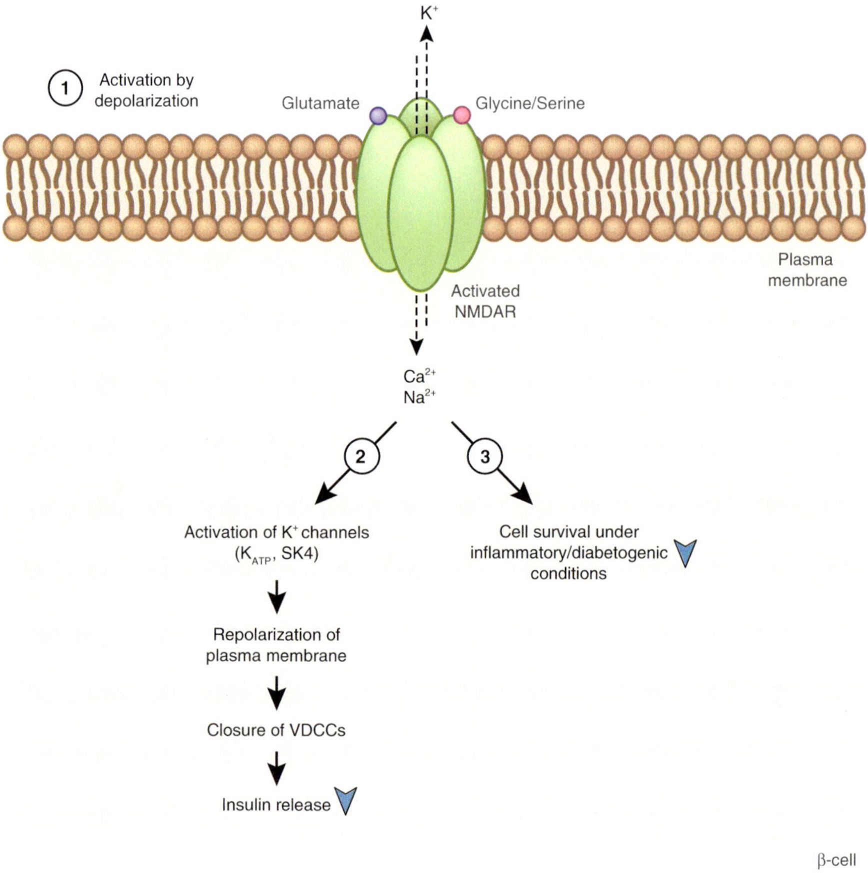

Fig. 7.1. Proposed NMDAR signaling in pancreatic beta cell. Membrane-bound ionotropic NMDAR is shown with its signaling pathway due to receptor activation. The steps (1–3) are explained in the main text. In brief, NMDAR is fully saturated with glutamate and mainly activated by the depolarization of β-cell membrane. NMDAR activity leads to an activation of K^+ channels (SK4 and especially K_{ATP} channels), thus to a repolarization of depolarized plasma membrane, closure of VDCCs and consequently to an inhibition of insulin release from β-cell. Furthermore, under inflammatory/diabetogenic conditions NMDAR activity contributes to islet cell death

channel opening and inhibit glucose-stimulated Ca^{2+} oscillations and insulin secretion in rat pancreatic β-cells [50]. Activation of K^+ channels, and thus K^+ efflux, leads to a repolarization of the depolarized plasma membrane, VDCC closure, and therefore reduced insulin release (Fig. 7.1., step 2). According to this model, NMDARs are part of a negative feedback mechanism that prevents excessive insulin release at high blood glucose concentrations.

7.3.3 Role of NMDARs in Beta Cell Death

Depending on their subcellular localization and subunit composition, neuronal NMDARs can induce either cell survival or cell death. Whereas activation of extra-synaptic NMDARs triggers cAMP response element binding protein (CREB) shut-off and activates cell death pathways, synaptic NMDARs promote the opposite [51]. Prior research suggested that extracellular signal-regulated kinase 1/2-dependent phosphorylation of the protein messenger Jacob signals the origin of NMDAR activation to the nucleus: activation of extrasynaptic NMDARs induces nuclear translocation of dephosphorylated Jacob leading to sustained dephosphorylation and thus inactivation of CREB, which is known to be associated with cell death [52]. Pancreatic NMDARs may resemble extrasynaptic NMDAR signaling because in pancreatic islets NMDAR activity decreases islet cell survival under inflammatory and diabetogenic conditions (Fig. 7.1., step 3). Treatment of isolated mouse and human pancreatic islets with inflammatory cytokines (TNF-α, IL-1β and IFN-γ) leads to increased islet cell death compared to untreated control islets. Further, inhibition of NMDAR activity attenuates islet cell death induced by these cytokines [40]. As low-grade inflammation appears to play a key role in the pathogenesis of type 2 diabetes mellitus (T2DM) [53], NMDAR signaling might play a role in triggering β-cell death and therefore progression to severe, insulin-dependent diabetes mellitus. Whether CREB shutoff and/or the protein messenger Jacob are involved in NMDAR-mediated islet cell death requires further investigation. It has also been suggested that NMDAR inhibition induces activation of calcineurin (a Ca^{2+}-regulated cytosolic phosphatase) and the PI3K/Akt signaling pathway, which is known to inhibit apoptosis and promote β-cell mass expansion in obesity [54].

7.4 NMDARs as Drug Targets for Diabetes Treatment

7.4.1 Demand for Beta Cell Protective Anti-diabetic Drugs

Currently, NMDARs are commonly-used drug targets for the treatment of several neurological diseases, including Alzheimer's disease and Parkinson's disease [37]. Considering their role in insulin secretion, blood glucose homeostasis and islet cell survival, pancreatic NMDARs are promising drug targets for adjunct treatment of T2DM and, possibly, type 1 diabetes mellitus (T1DM) as well [40, 55].

Diabetes mellitus is one of the largest global health problems. Over 415 million people are estimated to suffer from diabetes mellitus worldwide, with about 90–95 % affected by T2DM [56]. T2DM is characterized by peripheral insulin resistance and a progressive decline in functional β-cell mass, the latter due to β-cell dysfunction and β-cell death [57]. Initially, β-cells adapt to the diminished systemic tissue responses to insulin by increasing insulin secretion. However, over time, β-cells fail to compensate for peripheral insulin resistance and insulin secretion decreases. Both β-cell death and β-cell dedifferentiation have been proposed as the main causes of diabetic β-cell failure [58, 59]. In prediabetic individuals, the initial secretion defect is characterized by a weaker first phase of insulin secretion (i.e. the first 0–90 min following oral glucose uptake), and stronger second phase of insulin release (after 120 min following the oral glucose uptake) when compared to normoglycemic individuals [60]. Once the prediabetic individuals become diabetic (as defined by fasting plasma glucose concentrations equal or greater than 126 mg/dl), both phases of insulin secretion are often reduced [3, 56, 61].

Unfortunately, the most commonly prescribed, anti-diabetic drugs are not able to halt progressive β-cell dysfunction and β-cell death, or even to restore islet function to induce disease regression in individuals with T2DM [62]. Moreover, some anti-diabetic drugs, particularly insulin and sulfonylureas (the latter leading to a closure of the K_{ATP} channels in β-cells), can induce life-threatening hypoglycemia because serum insulin concentrations are not only increased under high, but also under low blood glucose conditions [63, 64]. Further, the biguanide metformin, which is the first-line drug for the treatment of T2DM because it effectively lowers blood glucose concentrations, can also lead to severe lactate acidosis and is therefore counterindicated during surgical interventions and in severe renal impairment [65]. More recently developed anti-diabetic drugs, such as inhibitors of the incretin-degrading dipeptidyl peptidase-4 (DPP-4), increase GSIS, but not basal insulin secretion, and thus avoid life-threatening hypoglycemic events [66]. However, the effects of DPP-4 inhibitors on blood glucose concentrations are similar to those of sulfonylureas [67]. Incretin-based drugs might help to slow down islet dysfunction and diabetes progression [68], even though a recent study indicated that incretins do not sustainably restore islet function [66]. The most recently developed anti-diabetic drugs are inhibitors of the sodium-glucose co-transporter-2 (SGLT-2). This transporter is present in the proximal tubules of the kidney and required for glucose reuptake. Therefore, inhibition of SGLT-2 facilitates excretion of glucose with the urine, thereby reducing blood glucose concentration and leading to weight loss [69]. Even though SGLT-2 inhibitors significantly lower blood glucose concentration and reduce cardiovascular mortality in patients with T2DM [70], they also introduce significant adverse events. These include urogenital tract infections and ketoacidosis, the latter being able to develop into a life-threatening event [71, 72]. Additionally, even these drugs become less effective at maintaining blood glucose homeostasis after a couple of years [73]. Therefore, the search continues for an ideal anti-diabetic drug that normalizes blood glucose concentration, assists with weight loss, avoids hypoglycemia, reduces cardiovascular mortality and promotes islet cell survival to delay, or even stop disease progression [74].

7.4.2 DXM and Derivatives in Preclinical Trials

Dextromethorphan (DXM) is an over-the-counter drug and noncompetitive NMDAR antagonist, which is already in use for the treatment of acute cough, pseudobulbar affect, diabetic neuropathic pain, and nonketotic hyperglycinemia

[75–78]. Following its absorption from the gut, DXM is quickly metabolized in the liver to its major active metabolite dextrorphan (DXO) by a cytochrome P450 enzyme [79]. DXO blocks the channel pore of the NMDAR in both its open and closed confirmation [80]. It was recently demonstrated that DXM and DXO have anti-diabetic effects both *in vitro* and *in vivo* [40, 55]. Blocking pancreatic NMDARs with DXM or DXO increases the insulin content of mouse pancreatic islets during long-term treatment of diabetic mice (Fig. 7.2., step 1), and enhances GSIS from mouse and human β-cells, thus reducing blood glucose concentration (Fig. 7.2., step 2). *In vitro* measurements of β-cell membrane potentials and cytoplasmic free Ca^{2+} in whole islets and β-cells of pancreatic slices further revealed that inhibition of NMDARs prolongs the amount of time that β-cells stay in the depolarized state with high cytosolic Ca^{2+} concentrations known to trigger insulin secretion [40]. Notably, basal insulin secretion remains largely unaffected upon NMDAR inhibition and the application of DXM does not introduce hypoglycemia even in the fasted state in mice and humans [40]. DXM and DXO treatment of pancreatic islets was furthermore shown to increase islet cell survival under inflammatory and diabetogenic conditions (Fig. 7.2., step 3). Specifically, treatment with DXO results in lower rates of human islet cell death induced by inflammatory cytokines (TNF-α, IL-1β and IFN-γ) compared to untreated human islets *in vitro*, and long-term application of DXM reduces islet cell death in diabetic mice *in vivo* [40].

7.4.3 DXM in Clinical Trials

Recent clinical trials revealed that a single oral dose of DXM improves glucose tolerance in individuals with T2DM and increases serum insulin concentrations, particularly during the first phase of an oral glucose tolerance test (OGTT) [40, 55] (Fig. 7.2., step 2). Furthermore, a low dose of DXM enhances the blood glucose lowering effect of the DPP-4 inhibitor sitagliptin and significantly increases serum insulin concentrations during the first phase of an OGTT in individuals with T2DM [55]. These clinical studies indicate that even low doses of DXM (30 or 60 mg) increase insulin release in humans with T2DM.

7.5 Outlook

Taken together, NMDAR antagonists could serve as novel anti-diabetic drugs due to their blood glucose lowering and islet cell protective effects. Treatment with NMDAR antagonists might be useful to delay or even prevent diabetes progression. Given their islet cell protective properties, NMDAR antagonists may also be useful in the treatment of autoimmune-mediated T1DM. Since DXM and DXO pass the BBB and interact with central NMDARs, they can induce central nervous side effects, such as fatigue and dizziness [40, 55]. The development of peripherally-restricted

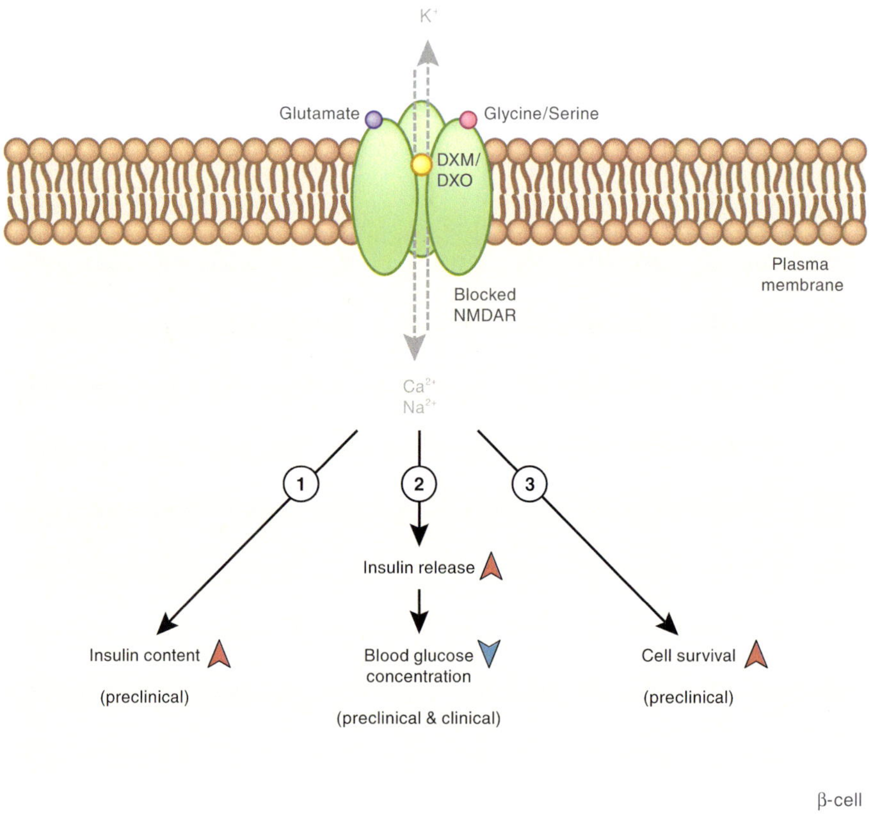

Fig. 7.2. Proposed consequences of NMDAR inhibition in beta cell. Membrane-bound ionotropic NMDAR is shown with the cellular consequences due to receptor inhibition. The steps (1–3) are explained in the main text. In brief, NMDAR is blocked by its antagonists DXM/DXO, which leads to an interrupted ion flow through the receptor. Consequences of this inhibition are an increase of β-cell insulin content, an enhanced insulin release from β-cell and thus a decrease of blood glucose concentration. Furthermore, islet cell survival increases due to NMDAR inhibition. In brackets: preclinical or clinical evidence of shown propositions

NMDAR antagonists and/or NMDAR antagonists with increased specificity towards pancreatic NMDAR subtypes would be useful to avoid CNS-related side effects. Further studies and long-term clinical trials are now needed to better understand the physiologic function and regulation of pancreatic NMDARs, and to assess the long-term effects of NMDAR inhibition on β-cell function and survival in T2DM patients in more detail. Finally, investigating the effects of NMDAR inhibition on cardiovascular endpoints (e.g. stroke and myocardial infarction), as well as diabetic neuropathies and retinopathies will be required and are of strong interest based on preclinical data indicating that DXM and DXO might have neuroprotective and arterioprotective effects [81–86].

References

1. Pandol SJ. The exocrine pancreas. San Rafael (CA); 2010.
2. Rorsman P, Braun M. Regulation of insulin secretion in human pancreatic islets. Annu Rev Physiol. 2013;75:155–79. Epub 2012/09/15
3. Rutter GA, Pullen TJ, Hodson DJ, Martinez-Sanchez A. Pancreatic beta-cell identity, glucose sensing and the control of insulin secretion. Biochem J. 2015;466(2):203–18. Epub 2015/02/24
4. Habegger KM, Heppner KM, Geary N, Bartness TJ, DiMarchi R, Tschop MH. The metabolic actions of glucagon revisited. Nat Rev Endocrinol. 2010;6(12):689–97. Epub 2010/10/20
5. Rodriguez-Diaz R, Caicedo A. Novel approaches to studying the role of innervation in the biology of pancreatic islets. Endocrinol Metab Clin North Am. 2013;42(1):39–56. Epub 2013/02/09
6. Eberhard D, Lammert E. The pancreatic beta-cell in the islet and organ community. Curr Opin Genet Dev. 2009;19(5):469–75. Epub 2009/08/29
7. Nussey S, Whitehead S. Endocrinology: an integrated approach. Oxford; 2001.
8. Alexandra E, Folias MH. Part VI pancreas, overview. In: MZ EL, editor. Metabolism of human diseases. Wien: Springer; 2014.
9. Ashcroft FM, Rorsman P. Diabetes mellitus and the beta cell: the last ten years. Cell. 2012;148(6):1160–71. Epub 2012/03/20
10. Gilon P, Ravier MA, Jonas JC, Henquin JC. Control mechanisms of the oscillations of insulin secretion in vitro and in vivo. Diabetes. 2002;51(Suppl 1):S144–51. Epub 2002/01/30
11. Gilon P, Chae HY, Rutter GA, Ravier MA. Calcium signaling in pancreatic beta-cells in health and in Type 2 diabetes. Cell Calcium. 2014;56(5):340–61. Epub 2014/09/23
12. Berggren PO, Larsson O. Ca2+ and pancreatic B-cell function. Biochem Soc Trans. 1994;22(1):12–8. Epub 1994/02/01
13. Satoh T. Molecular mechanisms for the regulation of insulin-stimulated glucose uptake by small guanosine triphosphatases in skeletal muscle and adipocytes. Int J Mol Sci. 2014;15(10):18677–92. Epub 2014/10/18
14. Soltani N, Qiu H, Aleksic M, Glinka Y, Zhao F, Liu R, et al. GABA exerts protective and regenerative effects on islet beta cells and reverses diabetes. Proc Natl Acad Sci U S A. 2011;108(28):11692–7. Epub 2011/06/29
15. Rodriguez-Diaz R, Dando R, Huang YA, Berggren PO, Roper SD, Caicedo A. Real-time detection of acetylcholine release from the human endocrine pancreas. Nat Protoc. 2012;7(6):1015–23. Epub 2012/05/05
16. Konstantinova I, Nikolova G, Ohara-Imaizumi M, Meda P, Kucera T, Zarbalis K, et al. EphA-Ephrin-A-mediated beta cell communication regulates insulin secretion from pancreatic islets. Cell. 2007;129(2):359–70. Epub 2007/04/24
17. Platt SR. The role of glutamate in central nervous system health and disease--a review. Vet J. 2007;173(2):278–86. Epub 2005/12/27
18. Di Cairano ES, Davalli AM, Perego L, Sala S, Sacchi VF, La Rosa S, et al. The Glial Glutamate Transporter 1 (GLT1) is expressed by Pancreatic beta-cells and prevents Glutamate-induced beta-cell death. J Biol Chem. 2011;286(16):14007–18.
19. Feldmann N, del Rio RM, Gjinovci A, Tamarit-Rodriguez J, Wollheim CB, Wiederkehr A. Reduction of plasma membrane glutamate transport potentiates insulin but not glucagon secretion in pancreatic islet cells. Mol Cell Endocrinol. 2011;338(1–2):46–57. Epub 2011/03/05
20. Gonoi T, Mizuno N, Inagaki N, Kuromi H, Seino Y, Miyazaki J, et al. Functional neuronal ionotropic glutamate receptors are expressed in the non-neuronal cell line MIN6. J Biol Chem. 1994;269(25):16989–92. Epub 1994/06/24
21. Inagaki N, Kuromi H, Gonoi T, Okamoto Y, Ishida H, Seino Y, et al. Expression and role of ionotropic glutamate receptors in pancreatic islet cells. FASEB Journal. 1995;9(8):686–91. Epub 1995/05/01
22. Molnar E, Varadi A, McIlhinney RA, Ashcroft SJ. Identification of functional ionotropic glutamate receptor proteins in pancreatic beta-cells and in islets of Langerhans. FEBS Lett. 1995;371(3):253–7. Epub 1995/09/11

23. Storto M, Capobianco L, Battaglia G, Molinaro G, Gradini R, Riozzi B, et al. Insulin secretion is controlled by mGlu5 metabotropic glutamate receptors. Mol Pharmacol. 2006;69(4):1234–41. Epub 2006/01/21
24. Otter S, Lammert E. Exciting times for pancreatic islets: Glutamate signaling in endocrine cells. Trends Endocrinol Metabolism. 2016;27(3):177–88. Epub 2016/01/08.
25. Maechler P, Wollheim CB. Mitochondrial glutamate acts as a messenger in glucose-induced insulin exocytosis. Nature. 1999;402(6762):685–9. Epub 1999/12/22
26. Gheni G, Ogura M, Iwasaki M, Yokoi N, Minami K, Nakayama Y, et al. Glutamate acts as a key signal linking glucose metabolism to Incretin/cAMP action to amplify insulin secretion. Cell Rep. 2014;9(2):661–73. Epub 2014/11/07
27. Vilsboll T, Holst JJ. Incretins, insulin secretion and Type 2 diabetes mellitus. Diabetologia. 2004;47(3):357–66. Epub 2004/02/18
28. Bertrand G, Puech R, Loubatieres-Mariani MM, Bockaert J. Glutamate stimulates insulin secretion and improves glucose tolerance in rats. Am J Physiol. 1995;269(3 Pt 1):E551–6. Epub 1995/09/01
29. Chevassus H, Renard E, Bertrand G, Mourand I, Puech R, Molinier N, et al. Effects of oral monosodium (L)-glutamate on insulin secretion and glucose tolerance in healthy volunteers. Br J Clin Pharmacol. 2002;53(6):641–3. Epub 2002/06/06
30. Cabrera O, Jacques-Silva MC, Speier S, Yang SN, Kohler M, Fachado A, et al. Glutamate is a positive autocrine signal for glucagon release. Cell Metab. 2008;7(6):545–54.
31. Huypens P, Ling Z, Pipeleers D, Schuit F. Glucagon receptors on human islet cells contribute to glucose competence of insulin release. Diabetologia. 2000;43(8):1012–9. Epub 2000/09/16
32. Omar B, Sorhede-Winzell M, Ahren B. Conditional glucagon receptor overexpression has multi-faceted consequences for beta-cell function. Metabolism. 2014;63(12):1568–76. Epub 2014/10/11
33. Wu ZY, Zhu LJ, Zou N, Bombek LK, Shao CY, Wang N, et al. AMPA receptors regulate exocytosis and insulin release in pancreatic beta cells. Traffic. 2012;13(8):1124–39. Epub 2012/05/01
34. Hawkins RA. The blood-brain barrier and glutamate. Am J Clin Nutr. 2009;90(3):867S–74S. Epub 2009/07/03
35. Gupta K, Hardingham GE, Chandran S. NMDA receptor-dependent glutamate excitotoxicity in human embryonic stem cell-derived neurons. Neurosci Lett. 2013;543:95–100. Epub 2013/03/23
36. Lammert E, Gu G, McLaughlin M, Brown D, Brekken R, Murtaugh LC, et al. Role of VEGF-A in vascularization of pancreatic islets. Curr Biol. 2003;13(12):1070–4. Epub 2003/06/20
37. Kalia LV, Kalia SK, Salter MW. NMDA receptors in clinical neurology: excitatory times ahead. Lancet Neurol. 2008;7(8):742–55. Epub 2008/07/19
38. Lutz TA, Meyer U. Amylin at the interface between metabolic and neurodegenerative disorders. Front Neurosci. 2015;9:216. Epub 2015/07/03
39. Eizirik DL, Cardozo AK, Cnop M. The role for endoplasmic reticulum stress in diabetes mellitus. Endocr Rev. 2008;29(1):42–61. Epub 2007/12/01
40. Marquard J, Otter S, Welters A, Stirban A, Fischer A, Eglinger J, et al. Characterization of pancreatic NMDA receptors as possible drug targets for diabetes treatment. Nat Med. 2015;21(4):363–72. Epub 2015/03/17
41. Ye L, Huang Y, Zhao L, Li Y, Sun L, Zhou Y, et al. IL-1beta and TNF-alpha induce neurotoxicity through glutamate production: a potential role for neuronal glutaminase. J Neurochem. 2013;125(6):897–908. Epub 2013/04/13
42. Paoletti P, Bellone C, Zhou Q. NMDA receptor subunit diversity: impact on receptor properties, synaptic plasticity and disease. Nat Rev Neurosci. 2013;14(6):383–400. Epub 2013/05/21
43. Garrino MG, Henquin JC. Adamantane derivatives: a new class of insulin secretagogues. Br J Pharmacol. 1987;90(3):583–91. Epub 1987/03/01
44. Ashcroft FM, Kerr AJ, Gibson JS, Williams BA. Amantadine and sparteine inhibit ATP-regulated K-currents in the insulin-secreting beta-cell line, HIT-T15. Br J Pharmacol. 1991;104(3):579–84. Epub 1991/11/01

45. Konrad D, Sobetzko D, Schmitt B, Schoenle EJ. Insulin-dependent diabetes mellitus induced by the antitussive agent dextromethorphan. Diabetologia. 2000;43(2):261–2. Epub 2001/02/07
46. Lechin F, van der Dijs B, Pardey-Maldonado B, Rivera JE, Lechin ME, Baez S. Amantadine reduces glucagon and enhances insulin secretion throughout the oral glucose tolerance test: central plus peripheral nervous system mechanisms. Diabetes Metab Syndr Obes. 2009;2:203–13. Epub 2009/01/01
47. Isaacson JS, Murphy GJ. Glutamate-mediated extrasynaptic inhibition: direct coupling of NMDA receptors to Ca(2+)-activated K+ channels. Neuron. 2001;31(6):1027–34. Epub 2001/10/03
48. Shen KZ, Johnson SW. Ca2+ influx through NMDA-gated channels activates ATP-sensitive K+ currents through a nitric oxide-cGMP pathway in subthalamic neurons. J Neurosci. 2010;30(5):1882–93. Epub 2010/02/05
49. Shen KZ, Yakhnitsa V, Munhall AC, Johnson SW. AMP kinase regulates K-ATP currents evoked by NMDA receptor stimulation in rat subthalamic nucleus neurons. Neuroscience. 2014;274:138–52. Epub 2014/05/31
50. Kaneko Y, Ishikawa T, Amano S, Nakayama K. Dual effect of nitric oxide on cytosolic Ca2+ concentration and insulin secretion in rat pancreatic beta-cells. Am J Physiol Cell Physiol. 2003;284(5):C1215–22. Epub 2003/01/17
51. Hardingham GE, Fukunaga Y, Bading H. Extrasynaptic NMDARs oppose synaptic NMDARs by triggering CREB shut-off and cell death pathways. Nat Neurosci. 2002;5(5):405–14. Epub 2002/04/16
52. Karpova A, Mikhaylova M, Bera S, Bar J, Reddy PP, Behnisch T, et al. Encoding and transducing the synaptic or extrasynaptic origin of NMDA receptor signals to the nucleus. Cell. 2013;152(5):1119–33. Epub 2013/03/05
53. Donath MY. Targeting inflammation in the treatment of type 2 diabetes: time to start. Nat Rev Drug Discov. 2014;13(6):465–76. Epub 2014/05/24
54. Wollheim CB, Maechler. *beta* cell glutamate receptor antagonists: novel oral antidiabetic drugs? Nat Med. 2015;21(4):310–1. Epub 2015/04/08
55. Marquard J, Stirban A, Schliess F, Sievers F, Welters A, Otter S, et al. Effects of dextromethorphan as add-on to sitagliptin on blood glucose and serum insulin concentrations in individuals with type 2 diabetes mellitus: a randomized, placebo-controlled, double-blinded, multiple crossover, single-dose clinical trial. Diabetes Obes Metab. 2016;18(1):100–3. Epub 2015/09/13
56. Diagnosis and classification of diabetes mellitus. Diabetes Care. 2014;37 Suppl 1:S81-90. Epub 2013/12/21
57. Stumvoll M, Goldstein BJ, van Haeften TW. Type 2 diabetes: principles of pathogenesis and therapy. Lancet. 2005;365(9467):1333–46. Epub 2005/04/13
58. Butler AE, Janson J, Bonner-Weir S, Ritzel R, Rizza RA, Butler PC. Beta-cell deficit and increased beta-cell apoptosis in humans with type 2 diabetes. Diabetes. 2003;52(1):102–10. Epub 2002/12/28
59. Talchai C, Xuan S, Lin HV, Sussel L, Accili D. Pancreatic beta cell dedifferentiation as a mechanism of diabetic beta cell failure. Cell. 2012;150(6):1223–34. Epub 2012/09/18
60. Mitrakou A, Kelley D, Mokan M, Veneman T, Pangburn T, Reilly J, et al. Role of reduced suppression of glucose production and diminished early insulin release in impaired glucose tolerance. N Engl J Med. 1992;326(1):22–9. Epub 1992/01/02
61. Nesher R, Cerasi E. Modeling phasic insulin release: immediate and time-dependent effects of glucose. Diabetes. 2002;51(Suppl 1):S53–9. Epub 2002/01/30
62. DeFronzo RA. Overview of newer agents: where treatment is going. Am J Med. 2010;123:S38–48.
63. Amiel SA, Dixon T, Mann R, Jameson K. Hypoglycaemia in type 2 diabetes. Diabetic Med. 2008;25(3):245–54. Epub 2008/01/25
64. Cryer PE. Mechanisms of hypoglycemia-associated autonomic failure in diabetes. N Engl J Med. 2013;369(4):362–72. Epub 2013/07/26
65. Bailey CJ, Tahrani AA, Barnett AH. Future glucose-lowering drugs for type 2 diabetes. Lancet Diabetes Endocrinol. 2016; Epub 2016/01/27

66. Retnakaran R, Kramer CK, Choi H, Swaminathan B, Zinman B. Liraglutide and the preservation of pancreatic beta-cell function in early type 2 diabetes: the LIBRA trial. Diabetes Care. 2014;37(12):3270–8. Epub 2014/09/25
67. Deacon CF, Lebovitz HE. Comparative review of dipeptidyl peptidase-4 inhibitors and sulphonylureas. Diabetes Obes Metab. 2016;18(4):333–47. Epub 2015/11/26
68. Kahn SE, Cooper ME, Del Prato S. Pathophysiology and treatment of type 2 diabetes: perspectives on the past, present, and future. Lancet. 2014;383(9922):1068–83. Epub 2013/12/10
69. Kalra S. Sodium Glucose Co-Transporter-2 (SGLT2) inhibitors: a review of their basic and clinical pharmacology. Diabetes Ther. 2014;5(2):355–66. Epub 2014/11/27
70. Zinman B, Lachin JM, Inzucchi SE. Empagliflozin, cardiovascular outcomes, and mortality in type 2 diabetes. N Engl J Med. 2016;374(11):1094. Epub 2016/03/18
71. Halimi S, Verges B. Adverse effects and safety of SGLT-2 inhibitors. Diabetes Metab. 2014;40(6 Suppl 1):S28–34. Epub 2015/01/03
72. Rosenstock J, Ferrannini E. Euglycemic diabetic ketoacidosis: a predictable, detectable, and preventable safety concern with SGLT2 inhibitors. Diabetes Care. 2015;38(9):1638–42. Epub 2015/08/22
73. Del Prato S, Nauck M, Duran-Garcia S, Maffei L, Rohwedder K, Theuerkauf A, et al. Long-term glycaemic response and tolerability of dapagliflozin versus a sulphonylurea as add-on therapy to metformin in patients with type 2 diabetes: 4-year data. Diabetes Obes Metab. 2015;17(6):581–90. Epub 2015/03/05
74. Tahrani AA, Bailey CJ, Del Prato S, Barnett AH. Management of type 2 diabetes: new and future developments in treatment. Lancet. 2011;378(9786):182–97.
75. Garnock-Jones KP. Dextromethorphan/quinidine: in pseudobulbar affect. CNS Drugs. 2011;25(5):435–45. Epub 2011/04/12
76. Nelson KA, Park KM, Robinovitz E, Tsigos C, Max MB. High-dose oral dextromethorphan versus placebo in painful diabetic neuropathy and postherpetic neuralgia. Neurology. 1997;48(5):1212–8. Epub 1997/05/01
77. Hamosh A, Maher JF, Bellus GA, Rasmussen SA, Johnston MV. Long-term use of high-dose benzoate and dextromethorphan for the treatment of nonketotic hyperglycinemia. J Pediatr. 1998;132(4):709–13.
78. Dicpinigaitis PV. Clinical perspective - cough: an unmet need. Curr Opin Pharmacol. 2015;22:24–8. Epub 2015/03/17
79. Nguyen L, Robson MJ, Healy JR, Scandinaro AL, Matsumoto RR. Involvement of sigma-1 receptors in the antidepressant-like effects of dextromethorphan. PLoS One. 2014;9(2):e89985. Epub 2014/03/04
80. LePage KT, Ishmael JE, Low CM, Traynelis SF, Murray TF. Differential binding properties of (3H)dextrorphan and (3H)MK-801 in heterologously expressed NMDA receptors. Neuropharmacology. 2005;49(1):1–16. Epub 2005/07/05
81. Tortella FC, Martin DA, Allot CP, Steel JA, Blackburn TP, Loveday BE, et al. Dextromethorphan attenuates post-ischemic hypoperfusion following incomplete global ischemia in the anesthetized rat. Brain Res. 1989;482(1):179–83. Epub 1989/03/13
82. Werling LL, Lauterbach EC, Calef U. Dextromethorphan as a potential neuroprotective agent with unique mechanisms of action. Neurologist. 2007;13(5):272–93. Epub 2007/09/13
83. Shin EJ, Bach JH, Lee SY, Kim JM, Lee J, Hong JS, et al. Neuropsychotoxic and neuroprotective potentials of dextromethorphan and its analogs. J Pharmacol Sci. 2011;116(2):137–48. Epub 2011/05/25
84. Liu SL, Li YH, Shi GY, Tang SH, Jiang SJ, Huang CW, et al. Dextromethorphan reduces oxidative stress and inhibits atherosclerosis and neointima formation in mice. Cardiovasc Res. 2009;82(1):161–9. Epub 2009/02/05
85. Wu TC, Chao CY, Lin SJ, Chen JW. Low-dose dextromethorphan, a NADPH oxidase inhibitor, reduces blood pressure and enhances vascular protection in experimental hypertension. PLoS One. 2012;7(9):e46067. Epub 2012/10/11
86. Liu et al. Treatment with dextromethorphan improves endothelial function, inflammation and oxidative stress in male heavy smokers. 2008. doi:10.1111/j.1538-7836.2008.03082.x

Chapter 8
NMDA Receptor in Bone

Yukio Yoneda

Abstract A view that L-glutamic acid (Glu) plays a role as an excitatory amino acid neurotransmitter through mechanisms relevant to activities of a variety of signaling machineries essential for the neurocrine at synapses in the brain is prevailing. Although expression of functional receptors is an absolute requirement for the glutamatergic signal input in the brain, recent molecular biological and pharmacological studies including ours give rise to a novel concept for Glu as an extracellular signal mediator in the autocrine and/or paracrine system in several non-neuronal tissues outside the brain. We have demonstrated functional expression of a variety of glutamatergic signaling machineries by bone-forming osteoblasts and mechanosensing osteocytes in bone, in addition to chondrocytes in cartilage, which are all derived from primitive mesenchymal stem cells in bone marrows. We could also detect functional expression of the cystine/Glu antiporter comprised of both xCT and 4F2hc subunits, rather than any other glutamatergic signaling machineries, by bone-resorbing osteoclasts believed to originate in hematopoietic stem cells. On the basis of these findings, we would propose a universal role of Glu as an extracellular signal mediator in the neurocrine, autocrine and paracrine systems in our body. Clinical aspect is also discussed on dietary Glu intake with a focus on possible benefits for the prophylaxis and/or treatment of osteoporosis.

Keywords Bone • Cartilage • Glutamate • Glutamate receptor • NMDA receptor • Osteoblast • Osteocyte • Osteoporosis • Chondrocyte • Mesenchymal stem cell

Abbreviations

AMPA	DL-α-amino-3-hydroxy-5-methylisoxasole-4-propionate
EAAC1	Excitatory amino acid carrier 1

Y. Yoneda, Ph.D. (✉)
Section of Prophylactic Pharmacology, Kanazawa University Venture Business Laboratory, Room 402, Kakuma-machi, Kanazawa, Ishikawa 920-1192, Japan
e-mail: yyoneda@p.kanazawa-u.ac.jp

K. Hashimoto (ed.), *The NMDA Receptors*, The Receptors 30,
DOI 10.1007/978-3-319-49795-2_8

EAAT	Excitatory amino acid transporter
GLAST	Glutamate aspartate transporter
GLT-1	Glutamate transporter-1
Glu	Glutamate
GluR	Glutamate receptor
GSH	Reduced glutathione
iGluR	Ionotropic glutamate receptor
KA	Kainate
M-CSF	Macrophage-colony stimulating factor
mGluR	Metabotropic glutamate receptor
MK-801	Dizocilpine
MNCs	Multinucleated cells
MSCs	Mesenchymal stem cells
NMDA	*N*-methyl-D-aspartate
NMDAR	*N*-methyl-D-aspartate receptor
RA	Rheumatoid arthritis
RANK	Receptor activator of NF-κB
RANKL	Receptor activator of NF-κB ligand
Runx2	Runt-related transcription factor-2
TRAP	Tartrate resistant acid phosphatase
VGLUT	Vesicular glutamate transporter

8.1 Background

In 1980s when L-glutamic acid (Glu) was not believed to be a neurotransmitter even in the brain, pioneering studies were done on the expression of possible glutamatergic signaling molecules in different peripheral tissues outside the brain in several independent laboratories. [^{3}H]Glu binding sites were found in rat adrenal [1], rat pituitary [2] and bovine pineal [3] glands with pharmacological profiles similar to those featuring in the brain, for instance, while both *N*-methyl-D-aspartic acid (NMDA) and Glu induced a rapid contraction of the guinea-pig ileum in a manner sensitive to antagonists [4–6]. Bronchial smooth muscle was shown to express a novel peripheral Glu receptor (GluR) subtype [7]. Recent progress of advanced technology led us to explore and confirm the functional expression of a variety of glutamatergic signaling machineries by cells in different peripheral tissues as well as the brain. These peripheral tissues included bone, testis, pancreas, adrenal, pituitary, pineal, taste buds, lung, hepatocyte, thymus, cerebral endothelium, megakaryocytes, keratinocytes, lymphocytes, platelets and heart as summarized in our previous review article [8], whereas this article highly focused on the expression by cells in bone.

8.2 Glutamatergic Signaling Machineries in Brain

In presynaptic glutamatergic neurons in the brain, intracellular Glu is condensed into synaptic vesicles by vesicular glutamate transporters (VGLUTs) located at vesicular surfaces toward exocytotic release into synaptic clefts in a Ca^{2+}-dependent manner upon stimuli, followed by activation of a variety of ionotropic (iGluRs) and metabotropic (mGluRs) receptor subtypes for this amino acid on plasma membranes in postsynaptic neurons for neurotransmission. Glutamatergic neurotransmission is rapidly arrested due to the incorporation mediated by excitatory amino acid transporters (EAATs) of extracellular Glu into adjacent cells including astrocytes. In astrocytes, Glu is converted into glutamine by the catalytic action of glutamine synthase, followed by efflux of glutamine into extracellular spaces. Extracellular glutamine is then incorporated into neighboring neurons for the conversion into Glu by glutaminase and Glu is condensed into synaptic vesicles as described above.

8.2.1 Glutamate Receptors for Signal Input

Membrane Glu receptors (GluRs) are endowed to transform extracellular signals carried by Glu into intracellular signals. These GluRs are nowadays categorized into two major subclasses, such as iGluRs and mGluRs receptors, on the basis of their differential intracellular signal transduction mechanisms and molecular homologies, with a pivotal role in mechanisms underlying neuronal plasticity such as learning and memory in the brain. Moreover, excessive overactivation of iGluRs is thought to participate in the molecular pathology of different neurodegenerative disorders such as Alzheimer's disease, Huntington's disease and amyotrophic lateral sclerosis, as well as neuronal cell death following a wide range of neurological insults including ischemia, trauma, hypoglycemia and epileptic seizures.

In contrast to other iGluR subtypes sensitive to DL-α-amino-3-hydroxy-5-methylisoxasole-4-propionate (AMPA) or kainate (KA), NMDA-sensitive iGluR subtype is highly permeable to Ca^{2+} along with sensitivity to blockade by Mg^{2+} at physiological concentrations in a voltage-dependent manner. Functional NMDA receptor (NMDAR) channels are usually orchestrated by tetrameric assemblies between the essential GluN1 subunit for the co-agonist glycine and one of four different GluN2 (A-D) subunits for the agonist Glu, in addition to one of two dominant negative GluN3 (A-B) subunits. Besides requirement for the agonist Glu and the co-agonist glycine, accordingly, expression of one of GluN2 subunits together with the essential GluN1 subunit is absolutely necessary for the orchestration of tetrameric functional NMDAR channels permeable to Ca^{2+} in any types of cells.

8.2.2 *Glutamate Transporters*

8.2.2.1 Signal Output

Three different isoforms of VGLUTs (VGLUT1 to VGLUT3) are recently characterized in the mechanism relevant to intracellular Glu transport for the condensation into synaptic vesicles at nerve terminals of glutamatergic neurons within the brain. Although the distribution of both VGLUT1 and VGLUT2 is highly complementary to each other in the brain, both isoforms are found on synaptic vesicles at excitatory synapses. Expression of either VGLUT1 or VGLUT2 would account for the exocytotic release of Glu in all known glutamatergic neurons, whereas VGLUT3 is expressed by a number of cells previously shown to release Glu through exocytosis including dopaminergic, GABAergic and serotonergic neurons as well as astroglia.

8.2.2.2 Signal Arrest

Glutamate transporters are supposed to be essential for the prevention of neurotoxicity as well as for the rapid arrest of signal transduction through lowering extracellular Glu levels in glutamatergic synapses. These transporters are classified into five different isoforms including glutamate aspartate transporter (GLAST) (EAAT1; excitatory amino acid transporter-1), glutamate transporter-1 (GLT-1) (EAAT2), excitatory amino acid carrier (EAAC1) (EAAT3), EAAT4, and EAAT5 to date. These isoforms display heterologous regional and cellular expression profiles. Both GLAST and GLT-1 isoforms are localized in astrocytes, for instance, with GLAST predominating in the cerebellum and GLT-1 in the cortex and forebrain, respectively. EAAC1 is localized in neurons throughout the CNS, while EAAT4 localization is largely restricted to cerebellar Purkinje cells. EAAT5 exclusively resides in the retina with specific location on photoreceptor and bipolar rod and cone cells.

8.2.2.3 Glutathione Synthesis

Sodium-independent, chloride-dependent high affinity Glu uptake system referred to as the cystine/Glu antiporter (=Xc- antiporter) is identified in plasma membranes of a variety of cells in different tissues. This antiporter is comprised of a heterodimeric assembly between the CD98 heavy chain 4F2hc, which is ubiquitously present in various tissues, and the xCT light chain, which determines the specificity for a substrate. The rate of cystine uptake by the cystine/Glu antiporter is a determinant for the regulation of intracellular reduced glutathione (GSH) levels. At high extracellular Glu concentrations, however, retrograde operation of the cystine/Glu antiporter occurs with extracellular Glu reversely taken up in exchange for intracellular cystine, which consequently leads to intracellular GSH depletion toward cell death mediated by oxidative stress in a manner irrelevant to the excitotoxicity of Glu.

8.3 Physiology of Bone

Bone is believed to undergo two distinct modes of the developmental process, such as intramembranous and endochondral ossifications, during embryonic skeletogenesis. The former process produces flat bones of the skull through the condensation of mesenchymal stem cells (MSCs), followed by differentiation of MSCs into the bone-forming osteoblasts and subsequent secretion of osteoid from those osteoblastic cells to form bone. In the later process, in contrast, a cartilage model is first formed by chondrocytes differentiated from MSCs, and then colonized MSCs differentiate into osteoblasts with deposition of osteoid on cartilage remnants in association with bone-resorbing osteoclasts. Mesenchymal stem cells located on periosteal surfaces and within bone marrow stroma are a source of osteoblasts for orchestration of bone during growth (=skeletogenesis) and maintenance of bone architecture in adulthood (=remodeling) in collaboration with osteoclasts. Both endocrine and paracrine (or autocrine) systems play a pivotal role in mechanisms underlying the embryonic skeletogenesis and adult bone remodeling. Impairment of these delicate balancing mechanisms leads to the pathogenesis as well as etiology of particular metabolic bone diseases including osteoporosis, Paget's disease and osteopetrosis. Bone-forming osteoblasts, bone-resorbing osteoclasts and mechanosensing osteocytes are all embedded within matrix to construct bone, while articulation is orchestrated by cartilages composed of chondrocytes at both distal ends of bones. In addition to bone and cartilage described below in this article, moreover, glutamatergic signaling system is functionally expressed by cells in a variety of peripheral tissues outside the brain as described elsewhere [8].

8.3.1 Mesenchymal Stem Cells

Mesenchymal stem cells are primitive pluripotent cells endowed to proliferate for self-renewal and to subsequently differentiate into a variety of progeny cell lineages, including osteoblasts, chondrocytes, adipocytes and myocytes. Moreover, evidence that MSCs differentiate into neuronal cells under particular *in vitro* conditions is now accumulating [9, 10]. In the endochondral ossification process, MSCs are absolutely responsible for skeletogenesis during development. For example, skeletal elements are originally differentiated from MSCs toward formation of a cartilaginous model during embryogenesis, which in turn leads to bone formation as endochondral ossification in the vertebral column and long bone. The cartilaginous rudiment, which is a tightly regulated area of both differentiation and maturation of chondrocytes, undergoes developmental growth for maturation. Within the cartilaginous rudiment, chondrocytes progressively differentiate through resting, proliferating, hypertrophic and calcifying stages, followed by mineralization of the cartilage matrix around the central region of the rudiment in the area of hypertrophic chondrocytes. Shortly after the mineralization process takes place, most

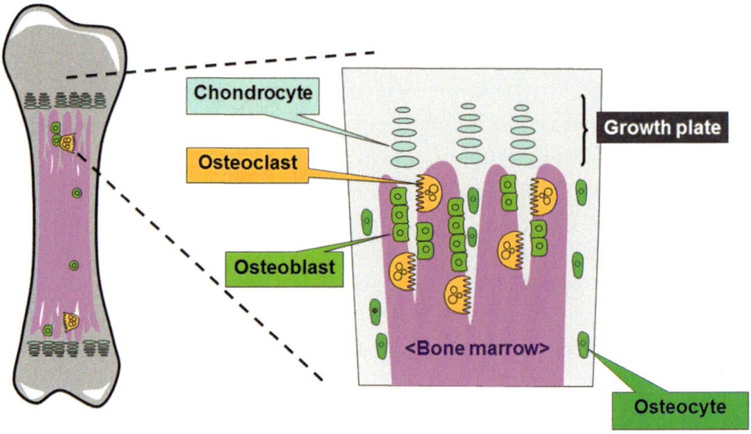

Fig. 8.1 Architecture of Long Bone. Long bone is constructed by cancellous area enriched of both osteoblasts and osteoclasts along with cortical area enriched of osteocytes, while cartilage is composed of chondrocytes to form articulation at the both distal ends of bone after closure of the growth plate

hypertrophic chondrocytes undergo sustained apoptosis. Upon apoptotic death of chondrocytes after mineralization, osteoblasts, osteoclasts and capillaries begin to invade the cartilage matrix to produce new bone responsible for the growth of endochondral bones (Fig. 8.1.).

8.3.2 Hematopoietic Stem Cells

Bone-resorbing osteoclasts are multinucleated cells (MNCs) derived from pluripotent hematopoietic stem cells shared with macrophage and dendritic cell lineages. Osteoclastogenesis is a multi-step process dependent on the presence of two important factors usually provided by cells of the osteoblastic lineage. Osteoblasts or bone marrow stromal osteoblastic precursors secrete macrophage-colony stimulating factor (M-CSF) along with expression of the cell-surface protein, receptor activator of NF-κB ligand (RANKL), which physically binds to its receptor (RANK) expressed on cellular surfaces of osteoclastic progenitors. Osteoblasts also express the soluble protein osteoprotegerin that acts as a decoy receptor for RANKL to inhibit osteoclastic differentiation. Fusion of mononuclear hematopoietic precursors is promoted after the intimate physical cellular contact between myeloid pre-osteoclastic precursors with RANK and osteoblasts (or stromal cells) with RANKL in association with M-CSF toward maturation to MNCs with an ability to resorb bone.

8.4 Glutamatergic Signaling Machineries in Bone

In addition to the aforementioned extracellular endocrine and paracrine factors, recent studies have raised the possible role of Glu as one of the endogenous factors used for intercellular communications in bone through a mechanism relevant to activation of NMDAR in osteoblasts and osteoclasts. For example, an NMDAR antagonist inhibited cell differentiation and bone-resorbing activities in cultured osteoclasts expressing both GluN1 and GluN2 subunits [11–13]. In these previous studies using bone marrow stromal cells for the growth of osteoclasts, however, the possibility that functional expression of different glutamatergic signaling molecules may be derived from stromal cells used for isolation, rather than osteoclasts themselves, is not ruled out.

8.4.1 Osteoblasts

Osteoblasts are derived from primitive MSCs in a state not terminally differentiated, while some osteoblasts ultimately become embedded in bone matrix to grow up to osteocytes with a property to gradually stop secreting osteoid. These osteocytes are localized in lacunae to form a network throughout mineralized bone tissue, communicating through gap junctions with each other as well as with surface osteoblasts. Other osteoblasts undergo apoptosis rather than becoming embedded in bone matrix for final differentiation into osteocytes. An alternative fate of cells of the osteoblast lineage is differentiation into the flattened bone-lining cells, which cover a substantial proportion of any bone surface.

In cultured human osteoblastic cell lines such as MG63 and SaOS-2 cells, Glu induced an elevation of intracellular free Ca^{2+} in a manner sensitive to the NMDA receptor antagonist dizocilpine (MK-801) [14]. Primary calvarial osteoblasts constitutively expressed both transcript and protein for both GluN1 and GluN2D subunits of NMDAR [15], and transcripts for GluA3 subunit of AMPA receptors and GluK1 and GluK2 subunits of KA receptors [16], while AMPA promoted exocytotic release of Glu from cultured calvarial osteoblasts in a Ca^{2+}-dependent manner [17]. In our studies, NMDAR activation predominantly modulated cell differentiation through a mechanism associated with upregulation of expression of runt-related transcription factor-2 (Runx2), which is an essential master transcription factor for osteoblast differentiation, during cellular maturation in calvarial osteoblasts [15]. Both transcript and corresponding protein expression was gradually increased for Runx2 in proportion to culture periods in rat calvarial osteoblasts, while the temporal upregulation was almost completely blocked by different antagonists for GluN1 and GluN2 subunits besides a channel blocker.

Prior stimulation of group I/group II mGluR led to reduction of NMDA-induced whole cell currents in cultured osteoblasts [18]. We have demonstrated constitutive expression of particular mGluRs (mGluR4 and/or mGluR8) negatively coupled to adenylyl cyclase to inhibit the formation of cAMP stimulated by forskolin in cultured osteoblasts [19]. However, no marked differences were shown in a variety of phenotypes including mandible and long bone sizes, morphology, trabeculation, regions of muscle attachment, resorption lacunae and areas of formation versus resorption of bone, compared with wild-type siblings in GLAST-null mice [20]. In contrast, mechanical loading down-regulated GLAST expression in osteocytes as revealed by immunohistochemistry [21]. One possible explanation for the lack of skeletal phenotypes in GLAST-null mice is lying on the heterogeneity of Glu transporters in osteoblasts as seen in the brain. Of EAAT isoforms cloned to date, in fact, both GLAST and GLT-1 isoforms were expressed in bone with regards to transcripts and proteins, without immunoreactive EAAC1 expression [21, 22].

8.4.2 *Osteocytes*

In contrast to osteoblasts described above, little attention has been paid to expression profiles of different glutamatergic signaling machineries in osteocytes, which are most abundant cells with longevity as a possible mechanical sensor in bone. Osteoblasts are endowed to differentiate into the third type of cells named osteocytes, which are surrounded by bone matrix to reside in lacunae. Osteocytes are equipped with long dendritic processes responsible for sensing mechanical forces, often referred to as mechanotransduction, toward mutual communication within osteocytes via interconnecting canaliculi [23].

In osteocytic MLO-Y4 cells with a much lower proliferation activity than osteoblastic MC3T3-E1 cells, transcript expression was seen for GluN1, GluN2D and GluN3B subunits of NMDAR, GluK2 and GluK5 subunits of KA receptor, mGluR8 isoform of mGluRs and VGLUT1 isoform of VGLUTs [24]. Sustained exposure to MK-801 led to a drastic increase in the number of cells with dendritic processes with a concomitant decrease in the number of cells without processes in cultured MLO-Y4 cells. Moreover, corresponding immunoreactivity was confirmed for both GluN1 and GluN2D subunits in osteocytes embedded within the cortical bone area in murine tibial sections. These immunohistochemical findings that both GluN1 and GluN2D subunits were abundant in numerous cells within cortical bone areas where osteocytes usually predominate [23], argue in favor of the possible expression of functional NMDAR channels composed of GluN1 and GluN2D subunits in osteocytes. Expression of VGLUT1 transcript also gives rise to an idea that the endogenous agonist Glu would be condensed into vesicular components in the cytoplasm for subsequent exocytotic release from osteocytes in a particular situation. From this point of view, it should be emphasized that both sympathetic and sensory nerve fibers innervate into bone together with glutamatergic innervation [25]. In osteoblasts, Glu could be supplied by glutamatergic nerve fibers as well as by osteoblasts themselves, in addition to blood circulation. Taken together, osteoblastic [16, 17] and osteocytic VGLUT would play a role as a machinery essential for the supply of extracellular Glu responsible for activation of particular GluR subtypes expressed by osteocytes and/or osteoblasts for glutamatergic signal input through the autocrine system in bone. Osteocytic NMDAR would play a role crucial for different functions, such as monitoring mechanical load and regulating bone remodeling, in bone.

8.4.3 Osteoclasts

Several independent lines of evidence indicated functional expression of NMDAR by osteoclasts, besides osteoblasts, as revealed by *in vitro* studies. In cultured osteoclasts prepared by isolation protocols using bone marrow stromal cells for the growth, the NMDAR antagonist MK-801 was shown to inhibit cellular differentiation. In cultured osteoclasts prepared by similar isolation procedures using stromal cells, Glu was accumulated into transcytotic vesicles by VGLUT1 for subsequent secretion [26]. To exclude the possible influence by bone marrow stromal cells with a phenotype of the osteoblastic lineage, however, we established accurate and reproducible isolation procedures for osteoclasts completely devoid of osteoblasts and bone marrow stromal cells [27]. In murine osteoclasts differentiated from bone marrow precursors in the presence of both recombinant M-CSF and RANKL, drastic upregulation was seen in transcript expression for all osteoclastic markers examined, but not for any osteoblastic markers, during culture. Accordingly, this novel cell isolation technique yielded populations of pure osteoclasts devoid of possible contamination with the osteoblastic cell lineage in contrast to osteoclasts previously prepared using bone marrow stromal cells. In these pre-osteoclasts and mature osteoclasts deprived of osteoblastic cells, surprisingly, no transcript expression was found for all GluRs and VGLUTs examined, and for GLAST, EAAC1 and EAAT5 isoforms of EAATs, except transcript for GLT-1 and EAAT4. Sequencing analysis on these amplified PCR products clearly confirmed the expression of each transcript for the corresponding signaling machineries. In mature osteoclasts, high immunoreactivity was detected for EAAT4, but not for GLT-1, isoform on Western blotting [27].

In contrast, constitutive transcript expression was found with both xCT and 4F2hc subunits for the functional heteromeric assembly to the cystine/Glu antiporter requited for the biosynthesis of the intracellular antioxidant GSH in osteoclasts defective of osteoblastic cells. Sustained exposure to Glu at high concentrations markedly inhibited the formation of mature osteoclastic MNCs positive for tartrate resistant acid phosphatase (TRAP) staining with a pavement-shape without affecting cellular viability. However, agonists for all iGluR and mGluR subtypes did not affect the number of TRAP-positive MNCs in cultured osteoclastic preparations devoid of osteoblastic cells. A significant decrease was seen in the endogenous levels of intracellular GSH in cultured osteoclastic cells defective of osteoblastic cells after sustained exposure to Glu at high concentrations [27]. Taken together, previous findings on functional expression by osteoclasts seem to be derived from artefactual presence of cells expressing functional glutamatergic signaling machineries with the osteoblastic phenotype, which were inevitably used for preparation of osteoclastic cells previously.

8.4.4 Chondrocytes

In order to focus on bone, this chapter provided less information on the functional expression of different glutamatergic signaling molecules in cartilage composed of chondrocytes than in bone. Nevertheless, we have demonstrated functional expression of particular glutamatergic signaling machineries by chondrocytes in cartilage. The group III mGluR agonist L-(1)-2-amino-4-phosphonobutyrate drastically inhibited chondral mineralization in a manner sensitive to an antagonist in cultured mouse embryonic metatarsals isolated before vascularization [28], for instance, while the addition of AMPA markedly evoked the release of endogenous Glu from cultured rat costal chondrocytes in a Ca^{2+}-dependent manner sensitive to potentiation by an AMPA receptor desensitization blocker [29]. Extracellular Glu was cleared up into intracellular locations through particular EAAT isoforms expressed by the rodent chondrocytes [30]. In addition, Glu cooperatively regulated cellular differentiation toward mineralization through a mechanism associated with the apoptosis after depletion of intracellular GSH due to the retrograde operation of the cystine/Glu antiporter [31], in addition to the aforementioned activation of group III mGluR, in chondrocytes. In contrast to different glutamatergic signaling molecules described above, we could not detect transcript expression of GluN1 subunit essential for orchestration of functional NMDAR channels by chondrocytes in cartilage. In addition, highly immunoreactive cells were detected for EAAC1 isoform in synovial membranes identified by their morphology and location, but neither GLAST nor GLT-1 isoform.

Accumulation of extracellular Glu could play a role in the pathogenesis of rheumatoid arthritis (RA) [32, 33]. A marked increase was seen in the endogenous levels of both Glu and aspartate in the synovial fluid from patients with arthritis [34], which is shown to induce increased edema volume and sensitized thermal hyperalgesia in arthritis model animals [35, 36]. In cultured synovial fibroblasts from RA model rats, markedly increased [^{3}H]Glu incorporation was invariably seen in a manner irrespective of the severity of the symptoms [32]. However, no significant changes were found in transcript expression of GluN2D, GluA3 and GluK2 subunits of iGluRs, mGluR8 isoform of mGluRs, GLAST, GLT-1 and EAAC1 isoforms of EAATs, and both xCT and 4F2hc subunits of the cystine/Glu antiporter between synovial tissues isolated from normal and RA animals. In synovial fibroblasts from RA model rats, by contrast, cellular proliferation activity was highly elevated in the presence of Glu in a fashion sensitive to EAAT inhibitors [32]. Taken together, extracellular Glu could promote cellular proliferation of synovial fibroblasts through a mechanism relevant to malfunction of particular EAAT isoforms during the progress of RA pathology, which consequently leads to the pathological destructions of both cartilage and bone in the articulation in association with several cytokines secreted from adjacent activated macrophages.

8.5 Clinical Aspect

To confirm the pathological and clinical significance of glutamatergic signals found in bone using the above-mentioned *in vitro* experimental protocols, we conducted *in vivo* experiments in ovariectomized female mice as an animal model of postmenopausal osteoporosis in humans. Daily intraperitoneal injection of Glu significantly prevented the reduction of bone mineral density in both tibia and femur along with increased Glu levels in bone marrows in ovariectomized mice. Moreover, the daily intraperitoneal administration of Glu not only prevented bone loss in the cancellous bone without affecting the cortical bone in ovariectomized mice on μCT analysis, but also protected alterations relevant to osteoporosis of different histomorphometric parameters. These included decreased bone volume/tissue volume (BV/TV) ratio, increased extent of eroded surface (ES/BS), increased number of osteoclasts on bone surfaces (Oc no) and increased extent of bone surface covered by osteoclasts (Oc surface) [27]. One favorable view is that exogenous Glu would promote osteoblastogenesis mediated by NMDAR activation in bone marrows, in addition to suppressing osteoclastogenesis through the retrograde cystine/Glu antiporter operation, toward prevention of bone loss in postmenopausal osteoporosis. Since dietary Glu is supposed to be unable to easily gain access to the circulating blood stream across intestinal mucosal cells [37], several devices are of course required for the effective delivery of dietary exogenous Glu to bone marrows toward beneficial prophylaxis and/or treatment of patients with osteoporosis in future studies.

8.6 Concluding Remarks

Our findings described above argue in favor of an idea that exogenous Glu loading is at least in part beneficial for the prophylaxis of bone loss in patients suffering from osteoporosis through mechanisms relevant to promoted osteoblastogenesis and suppressed osteoclastogenesis as shown in Fig. 8.2. It thus appears that Glu plays a pivotal role as a signal mediator in a variety of cells required for improvement of homeostasis in the neurocrine, autocrine and paracrine systems, rather than a roles as a simple excitatory amino acid neurotransmitter in the brain. Similar intercellular communications outside the brain are highly conceivable for diverse neurotransmitters other than Glu. Taken together, our findings about the new point of view are thus of a great interest and importance to wide spectra of basic and clinical researchers. This chapter dealt with the prime development of an innovative and novel interdisciplinary field as a scientific bridge between bone and brain biology. We are responsible for rapid and wide distribution of the information on advanced development of a novel interdisciplinary field to other scientists of different disciplines in discrete scientific fields. We are obliged to dispatch novel information important to scientists involved in bone and brain biology to elderly assistance in the world.

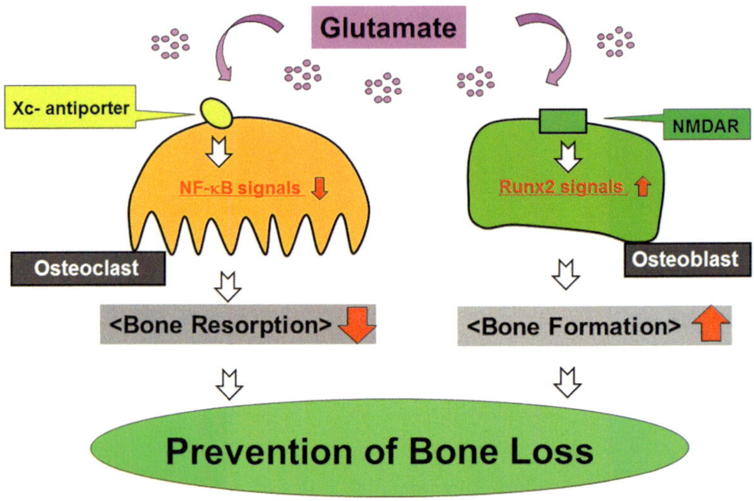

Fig. 8.2 Possible Glutamatergic Signaling in Bone. In bone-forming osteoblasts, extracellular Glu would promote osteoblastogenesis mediated by Runx2 signaling through activation of functional heteromeric NMDAR channels for Runx2 upregulation as an autocrine signal in bone marrows. In bone-resorbing osteoclasts, by contrast, Glu could suppress osteoclastogenesis mediated by NF-κB signaling through retrograde operation of the cystine/Glu antiporter for endogenous GSH depletion as a paracrine signal. The mechanism underlying the suppression of osteoclastogenesis after GSH depletion without cell death, however, remains to be elucidated. A strategy to increase Glu levels in bone marrows, accordingly, could be beneficial for the prophylaxis and/or treatment of bone loss in patients with osteoporosis in a particular situation

Acknowledgement The author is highly indebted to all colleagues listed in the references cited in this chapter for their excellent and enthusiastic efforts to support experimental studies for a period from 1999 to 2015 in Laboratory of Molecular Pharmacology, Kanazawa University Graduate School.

Conflict of interest: The author has no conflict of interest.

References

1. Yoneda Y, Ogita K. Localization of [^{3}H]glutamate binding sites in rat adrenal medulla. Brain Res. 1986a;383:387–91.
2. Yoneda Y, Ogita K. [^{3}H]Glutamate binding sites in the rat pituitary. Neurosci Res. 1986b;3:430–5.
3. Govitrapong P, Ebadi M, Murrin LC. Identification of a Cl^-/Ca^{2+}-dependent glutamate (quisqualate) binding site in bovine pineal organ. J Pineal Res. 1986;3:223–34.
4. Luzzi S, Zilletti L, Franchi-Micheli AM, Moroni F. Agonists, antagonists, and modulators of excitatory amino acid receptors in the guinea-pig myenteric plexus. Br J Pharmacol. 1988;95:1271–7.
5. Moroni F, Luzzi S, Franchi-Micheli S, Zilletti L. The presence of N-methyl-D-aspartate-type receptors for glutamic acid in the guinea-pig myenteric plexus. Neurosci Lett. 1986;68:57–62.

6. Shannon HE, Sawyer BD. Glutamate receptors of N-methyl-D-aspartate subtype in the myenteric plexus of guinea-pig ileum. J Pharmacol Exp Ther. 1989;251:518–23.
7. Aas P, Tanso R, Fonnum F. Stimulation of peripheral cholinergic nerves by glutamate indicates a new peripheral glutamate receptor. Eur J Pharmacol. 1989;164:93–102.
8. Hinoi E, Takarada T, Ueshima T, Tsuchihashi Y, Yoneda Y. Glutamate signaling in peripheral tissues. Eur J Biochem. 2004;271:1–13.
9. Sanchez-Ramos J, Song S, Cardozo-Pelaez F, Hazzi C, Stedeford T, Willing A, Freeman TB, Saporta S, Janssen W, Patel N, Cooper DR, Sanberg PR. Adult bone marrow stromal cells differentiate into neural cells in vitro. Exp Neurol. 2000;164:247–56.
10. Woodbury D, Schwarz EJ, Prockop DJ, Black IB. Adult rat and human bone marrow stromal cells differentiate into neurons. J Neurosci Res. 2000;61:364–70.
11. Chenu C, Serre CM, Raynal C, Burt-Pichat B, Delmas PD. Glutamate receptors are expressed by bone cells and are involved in bone resorption. Bone. 1998;22:295–9.
12. Mentaverri R, Kamel S, Wattel A, Prouillet C, Sevenet N, Petit JP, Tordjmann T, Brazier M. Regulation of bone resorption and osteoclast survival by nitric oxide: possible involvement of NMDA-receptor. J Cell Biochem. 2003;88:1145–56.
13. Peet NM, Grabowski PS, Laketic-Ljubojevic I, Skerry TM. The glutamate receptor antagonist MK801 modulates bone resorption in vitro by a mechanism predominantly involving osteoclast differentiation. FASEB J. 1999;13:2179–85.
14. Laketic-Ljubojevic I, Suva LJ, Maathuis FJ, Sanders D, Skerry TM. Functional characterization of N-methyl-D-aspartic acid-gated channels in bone cells. Bone. 1999;25:631–7.
15. Hinoi E, Fujimori S, Yoneda Y. Modulation of cellular differentiation by N-methyl-D-aspartate receptors in osteoblasts. FASEB J. 2003;17:1532–4.
16. Hinoi E, Fujimori S, Takemori A, Kurabayashi H, Nakamura Y, Yoneda Y. Demonstration of expression of mRNA for particular AMPA and kainate receptor subunits in immature and mature cultured rat calvarial osteoblasts. Brain Res. 2002a;943:112–6.
17. Hinoi E, Fujimori S, Takarada T, Taniura H, Yoneda Y. Facilitation of glutamate release by ionotropic glutamate receptors in osteoblasts. Biochem Biophys Res Commun. 2002b;297:452–8.
18. Gu Y, Publicover SJ. Expression of functional metabotropic glutamate receptors in primary cultured rat osteoblasts. Cross-talk with n-methyl-d-aspartate receptors. J Biol Chem. 2000;275:34252–9.
19. Hinoi E, Fujimori S, Nakamura Y, Yoneda Y. Group III metabotropic glutamate receptors in rat cultured calvarial osteoblasts. Biochem Biophys Res Commun. 2001;281:341–6.
20. Gray C, Marie H, Arora M, Tanaka K, Boyde A, Jones S, Attwell D. Glutamate does not play a major role in controlling bone growth. J Bone Miner Res. 2001;16:742–9.
21. Mason DJ, Suva LJ, Genever PG, Patton AJ, Steuckle S, Hillam RA, Skerry TM. Mechanically regulated expression of a neural glutamate transporter in bone: a role for excitatory amino acids as osteotropic agents? Bone. 1997;20:199–205.
22. Huggett J, Vaughan-Thomas A, Mason D. The open reading frame of the Na(+)-dependent glutamate transporter GLAST-1 is expressed in bone and a splice variant of this molecule is expressed in bone and brain. FEBS Lett. 2000;485:13–8.
23. Aarden EM, Burger EH, Nijweide PJ. Function of osteocytes in bone. J Cell Biochem. 1994;55:287–99.
24. Fujita H, Hinoi E, Nakatani E, Yamamoto T, Takarada T, Yoneda Y. Possible modulation of process extension by N-methyl-D-aspartate receptor expressed in osteocytic MLO-Y4 cells. J Pharmacol Sci. 2012;119:112–6.
25. Serre CM, Farlay D, Delmas PD, Chenu C. Evidence for a dense and intimate innervation of the bone tissue, including glutamate-containing fibers. Bone. 1999;25:623–9.
26. Morimoto R, Uehara S, Yatsushiro S, Juge N, Hua Z, Senoh S, Echigo N, Hayashi M, Mizoguchi T, Ninomiya T, Udagawa N, Omote H, Yamamoto A, Edwards RH, Moriyama Y. Secretion of L-glutamate from osteoclasts through transcytosis. EMBO J. 2006;25:4175–86.
27. Hinoi E, Takarada T, Uno K, Inoue M, Murafuji Y, Yoneda Y. Glutamate suppresses osteoclastogenesis through the cystine/glutamate antiporter. Am J Pathol. 2007;170:1277–90.

28. Wang L, Hinoi E, Takemori A, Takarada T, Yoneda Y. Abolition of chondral mineralization by group III metabotropic glutamate receptors expressed in rodent cartilage. Br J Pharmacol. 2005a;146:732–43.
29. Wang L, Hinoi E, Takemori A, Yoneda Y. Release of endogenous glutamate by AMPA receptors expressed in cultured rat costal chondrocytes. Biol Pharm Bull. 2005b;28:990–3.
30. Hinoi E, Wang L, Takemori A, Yoneda Y. Functional expression of particular isoforms of excitatory amino acid transporters by rodent cartilage. Biochem Pharmacol. 2005a;70:70–81.
31. Wang L, Hinoi E, Takemori A, Nakamichi N, Yoneda Y. Glutamate inhibits chondral mineralization through apoptotic cell death mediated by retrograde operation of the cystine/glutamate antiporter. J Biol Chem. 2006;281:24553–65.
32. Hinoi E, Ohashi R, Miyata S, Kato Y, Iemata M, Hojo H, Takarada T, Yoneda Y. Excitatory amino acid transporters expressed by synovial fibroblasts in rats with collagen-induced arthritis. Biochem Pharmacol. 2005b;70:1744–55.
33. Hinoi E, Yoneda Y. Possible involvement of glutamatergic signaling machineries in pathophysiology of rheumatoid arthritis. J Pharmacol Sci. 2011;116:248–56.
34. McNearney T, Speegle D, Lawand N, Lisse J, Westlund KN. Excitatory amino acid profiles of synovial fluid from patients with arthritis. J Rheumatol. 2000;27:739–45.
35. Lawand NB, McNearney T, Westlund KN. Amino acid release into the knee joint: key role in nociception and inflammation. Pain. 2000;86:69–74.
36. McNearney T, Baethge BA, Cao S, Alam R, Lisse JR, Westlund KN. Excitatory amino acids, TNF-alpha, and chemokine levels in synovial fluids of patients with active arthropathies. Clin Exp Immunol. 2004;137:621–7.
37. Riedijk MA, de Gast-Bakker DH, Wattimena JL, Van Goudoever JB. Splanchnic oxidation is the major metabolic fate of dietary glutamate in enterally fed preterm infants. Pediatr Res. 2007;62:468–73.